IMAGES, PERCEPTION, AND KNOWLEDGE

THE UNIVERSITY OF WESTERN ONTARIO
SERIES IN PHILOSOPHY OF SCIENCE

A SERIES OF BOOKS
ON PHILOSOPHY OF SCIENCE, METHODOLOGY,
AND EPISTEMOLOGY
PUBLISHED IN CONNECTION WITH
THE UNIVERSITY OF WESTERN ONTARIO
PHILOSOPHY OF SCIENCE PROGRAMME

Managing Editor

J. J. LEACH

Editorial Board

J. BUB, R. E. BUTTS, W. HARPER, J. HINTIKKA, D. J. HOCKNEY,
C. A. HOOKER, J. NICHOLAS, G. PEARCE

VOLUME 8

IMAGES, PERCEPTION, AND KNOWLEDGE

PAPERS DERIVING FROM AND RELATED TO THE PHILOSOPHY OF
SCIENCE WORKSHOP AT ONTARIO, CANADA, MAY 1974

Edited by

JOHN M. NICHOLAS

Dept. of Philosophy, University of Western Ontario, Ontario, Canada

D. REIDEL PUBLISHING COMPANY

DORDRECHT-HOLLAND / BOSTON-U.S.A.

Library of Congress Cataloging in Publication Data

Philosophy of Science Workshop, University of Western Ontario, 1974.
  Images, perception, and knowledge.

  (The University of Western Ontario series in philosophy of science ; v. 8)
  Bibliography: p.
  Includes index.
  1. Visual perception – Congresses. 2. Imagery (Psychology) – Congresses. 3. Cognition – Congresses.
  4. Psychology – Methodology. I. Nicholas, John M. II. Title.
  III. Series: University of Western Ontario. The University of Western Ontario series in philosophy of science ; v. 8.
  BF241.P46 1974   153.7   77-1264
  ISBN 90-277-0782-0
  ISBN 90-277-0783-9 pbk.

Published by D. Reidel Publishing Company,
P.O. Box 17, Dordrecht, Holland

Sold and distributed in the U.S.A., Canada, and Mexico
by D. Reidel Publishing Company, Inc.
Lincoln Building, 160 Old Derby Street, Hingham,
Mass. 02043, U.S.A.

All Rights Reserved
Copyright © 1977 by D. Reidel Publishing Company, Dordrecht, Holland
and copyrightholders as specified within
No part of the material protected by this copyright notice may be reproduced or
utilized in any form or by any means, electronic or mechanical,
including photocopying, recording or by any informational storage and
retrieval system, without written permission from the copyright owner

Printed in The Netherlands

TABLE OF CONTENTS

PREFACE	vii
ZENON W. PYLYSHYN / What the Mind's Eye Tells the Mind's Brain: A Critique of Mental Imagery	1
B. R. BUGELSKI / The Association of Images	37
ALLAN PAIVIO / Images, Propositions, and Knowledge	47
GEORGE W. BAYLOR and BERNARD RACINE / Mental Imagery and the Problems of Cognitive Representation: A Computer Simulation Approach	73
STEVEN ROSENBERG / The Separation and Integration of Related Semantic Information	95
AARON SLOMAN / Interactions Between Philosophy and Artificial Intelligence: The Role of Intuition and Non-Logical Reasoning in Intelligence	121
D. O. HEBB / Concerning Imagery	139
KARL H. PRIBRAM / Holonomy and Structure in the Organization of Perception	155
MARY HENLE / On the Distinction Between the Phenomenal and the Physical Object	187
D. W. HAMLYN / Unconscious Inference and Judgment in Perception	195
D. O. HEBB / To Know Your Own Mind	213
THOMAS NATSOULAS / The Subjective, Experiential Element in Perception	221
WILLIAM P. ALSTON / Can Psychology Do Without Private Data?	251
BIBLIOGRAPHY	291
INDEX	307

PREFACE

Despite the strictures of the extreme Behaviourists, psychologists have been taking an increasing interest in the development of theories concerning the 'mechanisms' *internal* to humans and animals which permit perceptual, memory, and problem solving behaviour. One consideration which has enormously stimulated an interest in theories of internal cognitive representation has been progress in the theory and the technology of computing machines, which has opened the promising prospect of computer simulation of human and animal psychological functions. What has developed is the possibility of constructing models of human psychology, realizing them in computer hardware, and testing the resultant machine performance against that of the human subject. A second consideration which helps motivate the construction of models of internal representation is the considerable advances in experimental and theoretical knowledge of the human brain understood from the neuro-anatomical view. The likely profit of adopting a narrowly Behaviourist methodology shrinks in the face of our growing, fine-grained knowledge of cerebral 'wetware'.

The purpose of this volume is selectively to exhibit some of the proposals concerning theories of internal representation which have been put forward in recent years. The area of central concern is the resurgence of interest in the role of imagery in cognition which has taken place in the last fifteen years. Despite its being banished from scientific psychology in the earlier part of this century, most notoriously in the context of the development of a Behaviourist hegemony in North America, and under the impact of the results of the Würzburg school concerning imageless thinking, too many interesting experimental results concerning the capacities of imagery in long-term memory tasks appeared during the 1960's for it to be left beyond the pale. Nor has interest been restricted to cataloguing experimental curiosities, for it has been proposed that image concepts have a deep explanatory role in cognitive theory, in particular, with respect to memory. Paivio (see, for example, his 'Images, Propositions, and Knowledge', below) has proposed a Dual Coding Theory which envisages a system used in human information representation which has distinct linguistic and imaginal modalities of storage. On this view the imagery system is seen as involving parallel-processing informational output while the verbal system is essentially sequential. The former is understood as being more closely linked with analogue forms of representation rather than digital.

The challenge of the role of imagery, thus experimentally and theoretically entrenched, has been taken up from different viewpoints. For example, Pylyshyn, in his influential 'What the Mind's Eye Tells the Mind's Brain', argues that we need not be constrained by the fact that there is introspectable imagery accompanying various intellectual and perceptual processes to suppose that imagery *per se* or even anything similar to it serves important information-processing purposes in the underlying cognitive representations. Instead, he draws attention to a form of internal representation which has been termed *conceptual* and *propositional*, and which is itself neither verbal nor imaginal — a sort of *interlingua*. Baylor and Racine, in 'Mental Imagery and the Problems of Cognitive Representation', attack the problem of simulating certain functions of imagery in problem-solving situations which involve mental figural transformations. Broadly speaking, they plump for a digital, discrete point of view concerning the underlying processes. Rosenberg's 'The Separation and Integration of Related Semantic Information' gives an account of certain suggestive experimental results which confirm a model of a *single* semantic representation which is independent of the modality of presentation of information, that is, whether it is imaginal or verbal. Sloman, in his 'Interactions between Philosophy and Artificial Intelligence', draws a distinction between what he calls Analogical and Fregean modes of representation, and proposes that the former has certain advantages over the latter for an intelligent robot. On the way he discusses the possibility of generalizing certain logical concepts to the non-linguistic forms of representation. In sum, the climate has changed for the image concept, at the very least permitting the calling into question of a globally verbal interpretation of thinking, and perhaps even the reappearance in a new guise of some very traditional hypotheses in this area, as in Bugelski's 'The Association of Images'.

Both Hebb's 'Concerning Imagery', and Pribram's 'Holonomy and Structure in the Organization of Perception' provide synthetic approaches to cognitive problems, which go so far as to propose physiological mechanisms which may underlie imagery. Hebb suggests that his theory of cell assemblies can provide the framework for understanding various kinds of imagery. Pribram borrows the concept of the hologram from physical optics, and gives an account of the cerebral underpinnings of imagery in perception and memory in terms of it. In so doing, he takes the bold step of shifting emphasis from neuronal impulses to the graded potentials which are developed between neurons at synaptic junctions.

The psychologist's espousal of theories of internal representation has many methodological presuppositions and consequences. Mary Henle in 'On the Distinction between the Phenomenal and the Physical Object' takes the opportunity to press for the recognition of the distinction in her title. She especially emphasizes that recognition does not force upon us many of the paradoxes which, outside the Gestalt school, it has been taken to do. Hamlyn's 'Unconscious Inference . . . ' provides critical discussion of one of the proposed ways by which internal constructs may be linked with a knowledge of the external world. Hebb's 'To Know Your Own Mind' offers a methodological blast against the scientific validity of introspection, while Alston, in 'Can Psychology Do Without Private Data?', attempts to clear the ground for the legitimate consideration of private data. This is clearly a matter of considerable delicacy for researchers working in perceptual psychology, and has a special significance for those who are forced to rely on a subject's protocols for guidance about their phenomenal imagery in cognitive tasks. Natsoulas, in 'The Subjective, Experimental Element in Perception', provides a valuable discussion of both the psychological and philosophical literature in this area from the viewpoint of the psychologist.

The papers by Bugelski, Paivio, Baylor and Racine, Rosenberg, Hebb ('To Know Your Own Mind'), Pribram, Henle, and Hamlyn derive from presentations at the 1974 Philosophy of Science Workshop of the Department of Philosophy at The University of Western Ontario. I wish to take the opportunity here to express thanks on behalf of the Department and the University to the Canada Council for its generous conference grant to that workshop. I wish also to thank U.W.O.'s Departments of Psychology, and History of Medicine and Science for their support. I thank Allan Paivio and Zenon Pylyshyn for some very valuable discussions about the complexities of this area of research. Thanks go too to Robert Priest for his assistance with a number of the diagrams in this volume.

JOHN M. NICHOLAS

ZENON W. PYLYSHYN

WHAT THE MIND'S EYE TELLS THE MIND'S BRAIN: A CRITIQUE OF MENTAL IMAGERY*

This paper presents a critique of contemporary research which uses the notion of a *mental image* as a theoretical construct to describe one form of memory representation. It is argued that an adequate characterization of 'what we know' requires that we posit abstract mental structures to which we do not have conscious access and which are essentially *conceptual* and *propositional*, rather than sensory or pictorial, in nature. Such representations are more accurately referred to as symbolic descriptions than as images in the usual sense. Implications of using an imagery vocabulary are examined, and it is argued that the picture metaphor underlying recent theoretical discussions is seriously misleading — especially as it suggests that the image is an entity to be perceived. The relative merits of several alternative modes of representation (propositions, data structures, and procedures) are discussed. The final section is a more speculative discussion of the nature of the representation which may be involved when people 'use' visual images.

Cognitive psychology is concerned with two types of questions: What do we know? and How do we acquire and use this knowledge? The first type of question, to which this paper is primarily addressed, concerns itself with what might be called the problem of cognitive representation. It attempts to answer the question What is stored? by describing the form in which our knowledge or model of the world is represented in the mind. The second type of question attempts to deal more directly with certain limited-capacity psychological processes which create and manipulate the representations and generate appropriate behavior from them (such as questions of limited attention and memory and 'psychological cost' factors limiting performance).

The problem of cognitive representation has been approached by a wide variety of paths in the last half century. It is not the purpose of this paper to review this spectrum of alternative representations. We concern ourselves with one particular proposal — that cognitive representations take the form of mental images — which has recently exploded into fashion and with a number of alternatives which derive primarily from computer-simulation work.

John M. Nicholas (ed.), Images, Perception, and Knowledge, 1–36.
First published in: Psychological Bulletin 80 (1973), 1–24.
Copyright © 1973 by the American Psychological Association.
Reprinted by permission.

1. MENTAL IMAGES

After almost 50 years of dedicated avoidance, mental imagery appears to be once again at the center of interest in many areas of psychology (Arnheim, 1969; Bower, 1972; Bugelski, 1970; Hebb, 1968; Holt, 1964; Horowitz, 1970; Paivio, 1969, 1971; Reese (1970); Richardson, 1969; Segal, 1971; Sheehan, 1972), although it is not without its detractors (Brainerd, 1971; Brown, 1958; Gibson, 1966; Neisser, 1972). It has returned not only as a phenomenon to be investigated but as an explanatory construct in cognitive psychology. It is of some interest, therefore, to ask whether anything new has been learned about mental images in the last decade and whether the critiques widespread early in the century have been satisfactorily countered.

Any analysis in the nature and role of imagery is fraught with difficulty. The concept itself proves to be difficult to pin down. Is a visual image like some conceivable picture? If not, then in what way must it differ? If it is like a picture in some ways, then must it always be a picture of some specific instance, or can it be generic (if such a notion is intelligible)? Could it, for example, represent abstract relations or must the relations in the image be of an iconic or geometric variety? Is an entire image available at once — as a spatially parellel static picture — or do parts of it come and go? If parts can be added and deleted at will, must such parts be pictorial segments (e.g., geometrically definable pieces or sensory attributes such as color) or can they be more abstract aspects? Could one, for example, conceive of two images of the identical chessboard with one image containing the relation 'is attacked by' and the other not containing it? If so, then in what sense could such a relation be said to be 'in the image'? Must images in some important sense be modality specific, as implied by such phrases as visual image, auditory image, etc.? And finally, must images always be conscious? Can one, for example, make intelligible the notion of an unconscious visual image?

The tumultuous history of the concept of imagery in both philosophy and psychology attests to the difficulties which such questions have raised in the minds of centuries of scholars. While contemporary psychologists have attempted to narrow the scope of the concept by operational definitions and multiple empirical and theoretical underpinnings, it is not clear that they have resolved the major *conceptual* ambiguities, circularities, and Rylean 'category mistakes' which have plagued the notion in the past. In this paper we do not attempt a philosophical analysis of the concept since that would take us too far afield from our primary objective, which is to analyze the role of imagery as an explanatory construct in cognitive psychology. In attempting this, however, it is impossible to avoid pointing out some of the conceptual problems implicit in contemporary uses of the term 'imagery' to explain certain findings in psychology.

For the sake of avoiding any misinterpretations of the remarks in the remainder of this paper, it should be stressed that the existence of the experience of images cannot be questioned. Imagery is a pervasive form of experience and is clearly of utmost importance to humans. We cannot speak of consciousness without, at the same time, implicating the existence of images. Such experiences are not in question here. Nor, in fact, is the status of imagery either as object of study (i.e., as dependent variable) or as scientific evidence being challenged. There are many areas where the alternative to cautious acceptance of reports of imagery is the rejection of a whole area of inquiry (e.g., Hebb, 1960; Holt, 1964). Furthermore, the extensive experimental investigations of imagery in the last decade (exhaustively surveyed in Pavio, 1971) have been of unquestionable value in breaking through the earlier oppressive structures on what phenomena ought to be studied. The empirical regularities demonstrated are of high reliability and wide interest, both from a scientific and practical point of view. None of these empirical results are questioned. The main question that is raised is whether the concept of image can be used as a primitive explanatory construct (i.e., one not requiring further reduction) in psychological theories of cognition. A second and equally important issue is whether the commonsense understanding of this term contains misleading implications which carry over undetected into psychological theories.

1.1. *Information and Experience*

While most psychologists are willing to concede that not all important psychological processes and structures are available to conscious inspection, it is not generally recognized that the converse may also hold: that what is available to conscious inspection may not be what plays the important causal role in psychological processes. This is not to imply that when a subject says that he does such-and-such a mental operation that he may be dishonest in his reports nor even that he himself is misled as to what he "actually" does. This is not the issue. If someone asks me how many windows there are in my house and asks me to report how I went about answering the question, this report (subject to the usual methodological precautions) may be taken as an accurate report of what I *experienced* doing. The only trouble is that I must give such a report in the only language I have available for describing my awarenesses. A description in such a language may be entirely inadequate in meeting what Chomsky (1964) has referred to as 'explanatory adequacy'. The description may even fail to be mechanistic insofar as it may use terms such as 'want', 'guess', 'notice', or mind's eye', which, while they may faithfully reflect ones experience of doing the task, are inadequate as *primitive constructs* since they themselves cry out for a reduction to mechanistic terms.

Even if the description does not contain inadmissible terms, there is no guarantee that it will make sense as an *explanatory* scientific theory. An explanatory theory must meet different criteria of adequacy than would be demanded of an informal *descriptive* account. It must first be free of conceptual difficulties and internal contradictions. Then it must be shown to be capable of providing a mechanistic explanation for the widest possible domain of empirical evidence in a manner which reveals the most general principles involved. An explanatory account must ultimately appeal to universal mechanistic principles. Just because we *know* that we use certain mnemonic strategies, or that we say certain things to ourselves, or that we 'see' certain objects in our 'mind's eye' or 'hear' ourselves rehearsing a series of numbers, etc., we cannot assume that the contents of such subjective knowledge can be identified with the kind of information-processing procedures which will go into an explanatory theory.

Perhaps these remarks can be made clearer if we sharply distinguish two senses of cognitive verbs like 'see', 'hear', or 'image'. In one sense, they refer to information – processing functions – to the reception of information through the visual or auditory systems and the subsequent transformation or encoding of this information in interaction with stored information. In the case of imagery it may perhaps refer to the activation, retrieval, or reorganization of such information. Another very different sense of these terms is implied, however, when they are used to designate the conscious experience which may accompany such functions, that is, the conscious experience of seeing, hearing, or imaging (or examining an image in the 'mind's eye'). In this sense of the word 'image' one might claim that the image can be examined through introspection. Clearly, however, the information-processing function itself cannot. It is important to inquire whether the experience of imaging can reveal important properties of the information-processing functions or of the mental representation of information on which these processes operate. But we must not assume in advance that such observation will reveal the content of the mental representation. Not only does such observation present serious methodological hazards, it is not prima facie an observation of the functional representation (i.e., one that figures in the human information-processing function).

In discussing the question of introspective knowledge, Natsoulas (1970) warns,

> It may turn out that, even though a useful informational relation (concomitant variation) exists between the contents of our awareness and the properties of mental episodes, they do not have the intrinsic properties which we take them to have (p. 91).

The recent literature on imagery abounds in examples which reveal that the investigator tacitly assumes that what is functional in cognition is available to introspection. Consider, for example, the widely held view (the so-called dual-code model) that the form of mental representations is either verbal or imaginal. This partition between two concrete modes has its roots in the persuasive fact that the only way in which we clearly experience our memory of cognitive events is through some form of sensory-motor image (including articulatory and acoustical images of words). Thus, for example, in a revival of a position associated with Berkeley and Hume, Bugelski (1970) questions 'whether there is such a thing as an abstract thought or abstraction (p. 1006)'. The basis of his doubts are his experiments, using the Kent-Rosanoff Word Association Test, in which he finds:

> If you say FLOWER, a categorical term, the subjects think of daisies or roses, and highly specific daisies or roses... If you say DEMOCRACY, they report a variety of imagery, practically none of which refers to governmental operations. Government by the people becomes an image of a crowd at a political rally (p. 1006).

Drawing conclusions about the nature of cognitive representations from reports of experiences evoked by words may appear some-what far-fetched (after all, what else could a subject report having experienced — other than images of objects or of other words?) until we examine the context in which Bugelski and his colleagues are working. The thrust of Bugelski's paper is to show the inadequacy of theories of learning and memory which rely exclusively on postulating associations among words. From this excess he adopts another equally untenable position (which is nowhere stated explicitly): that all learning and memory — and indeed all of cognition — takes place exclusively through the medium of either words or images. In fact, it appears that most modern psychologists working on imagery and learning have succumbed to the assumption that there are no forms of mental representation other than words and images. Thus Bugelski (1970) argues that the use of young deaf subjects who 'have no language' provides an ideal test for the existence of imagery. He asserts, 'If they truly have no speech or verbal capacity and can learn certain kinds of materials, for example, picture paired-associates, the conclusion that imagery was being used seems logically determined (p. 1004)'. It is logically determined, of course, only if we accept that images and words exhaust the available forms of mental representation.

Similarly, Paivio (1969) pits his defense of imagery against the word-association approach, arguing that ' ... one can respond verbally to pictures as well as to words and so, by analogy, one's verbal response could just as logically be mediated by a "mental picture" as by "mental words" (p. 242)'. The parallel does indeed hold: whatever arguments may be marshalled in favor of mental words as mediators can be used equally well to support mental pictures. Thus adding images to the repertoire of mediators represents a logical extension of mediational accounts. What is unsatisfactory about this extension, however, is that no consideration is given to the possibility that cognition may be 'mediated' by something quite different from either pictures or words, different in fact from anything which can be observed either from within or from without.

In spite of the prevalence of such views, a number of theorists, particularly those working in the information-processing tradition, have found it necessary to postulate forms of representation which differ radically from the form in which such information is presented to the senses or the way in which it is subjectively experienced. For example, many models of attention (e.g., Triesman, 1964) and memory (e.g., Norman, 1968) as well as all analysis-by-synthesis models (e.g., Neisser, 1967) require that representations differing from both the sensory pattern and the name or verbal description be available at some stage in the process. In the same spirit, the author argued some years ago (Pylyshyn, 1963) against the view that representations in short-term memory went directly from a visual to an auditory form as the information was 'read' off the image. Instead, a coding continuum was proposed. Sperling (1963) also found evidence against the two-stage view. He showed that while information sufficient to identify a letter could be extracted from a display in about 10 milliseconds, it took over 300 milliseconds to name a letter. Sperling developed a model in which information between these two stages (i.e., visual and verbal) was held in a 'recognition buffer memory'. Not only is the information in this buffer neither in a visual nor an auditory form, it is not in any form which *could* be made conscious. Sperling (1967) comments,

This makes it indeed a mysterious component; it cannot be observed directly from within or from without! However, this inaccessibility should not surprise us. It is axiomatic that in any system which examines itself there ultimately must be some part of the mechanism which is inaccessible to examination from within (p. 292).

This is indeed true; only perhaps the surprise ought to be that *any* of it should be accessible from within. One scarcely expects brain processes to be available to introspective examination, so why should one expect functional information to be thus accessible?

But the need to postulate a more abstract representation — one which resembles neither pictures nor words and is not accessible to subjective experience — is unavoidable. As long as we recognize that people can go from mental pictures to mental words or vice versa, we are forced to conclude that there must be a representation (which is more abstract and not available to conscious experience) which encompasses both. There must, in other words, be some common format or interlingua. The problem is dramatized if we persist in using the common but utterly misleading metaphor of the 'mind's eye', for then we have to account for the form of representation in the 'mind's eye's mind' which clearly is not accessible to introspection.

Any attempt to bypass this difficulty by positing a direct associative link between a mental picture and a mental word meets with other difficulties. There are an infinite number of pictures to which a particular word applies. For example, there are an infinite number of rectangles of various shapes, sizes, colors, orientations, etc. When the mental word 'rectangle' is elicited by the mental picture of a rectangle, it cannot be by virtue of an associative link between the two, since this would require that we postulate an infinite number of such links (one for each possible picture). The mental word 'rectangle' is at best a response to what all the pictures of rectangles have in common, namely, their 'rectangleness'. The problem arises because cognition must deal with pattern *types* and not *tokens*, and there is no limit to the variety of tokens corresponding to each type. Thus the relationship cannot be described by a finite list of associated picture-word tokens.

1.2 Propositions and Appearances

Although Bower (1972) appears to recognize the existence of such problems to the extent of a passing admission that people do have concepts — and that these are different from either words or pictures — he nevertheless proceeds to develop an argument for a dual-code model of memory of the type earlier advocated by Paivio (1969). His arguments are of interest because they shed some light on the nature of some of the conceptual difficulties to which such models give rise. In his account, Bower (1972) make a distinction between memory for appearances (*how* things look) and memory for what we call facts (*what* things resemble). From this he argues,

> This distinction between how something looked and what is looked like runs parallel to the distinction between cognitive memories, namely, images versus propositional memory. That is, we remember appearances in imagery, and we also remember propositions . . . the difference . . . is the same as the difference between a sighted versus a blind person's knowledge of the visual world. In the auditory domain, it is the difference between my knowledge of how an orchestra symphony sounds and how I might try to describe it to a deaf person (p. 52).

Not only is this dichotomy not exhaustive, it also gives a misleading account of what it means to 'know' such things as 'how the orchestra sounds'. To make this clear, we need to draw some further distinctions. First, we must distinguish between the subjective experience of recalling the sound and the functional information which enables us to, say, make judgments of relative similarity of two instruments. The term *appearance* is surely meant to refer only to the former. An appropriately programmed computer could, no doubt, be made to produce similarity judgments among sounds but we might still be reluctant to describe it as 'experiencing' the sound and therefore of being able to recall its appearance. It would, in general, even be unreasonable to require the computer to store the equivalent of a sound-recording trace of the stimulus, since this would soon tax the storage capacity of any machine. So even in this nonexperiential sense of appearance, it is still the case that the computer would not have access to the original pattern of stimulation when it made its similarity judgment. Thus there is no reasonable sense in which the computer would use the *appearance* of the original sound to make its judgment. While a person might *experience* the appearance of the sound, it would similarly be unreasonable to suppose that he can make the similarity judgment because he has stored the original pattern of sensory stimulation (we shall return to this point later.)

Second, when Bower and others speak of nonimaginal memory as propositional, they imply the storage of actual utterances. But one must be careful not to equate a proposition with a string of words. A proposition is what a string of words may assert. A proposition is either true or false; the string of words is neither. A proposition may be asserted by any number of strings of words, in any language and in any modality. Furthermore, in the sense in which we use it when we speak of 'propositional knowledge', it may involve no words of any kind. Thus when I look at the table in front of me, I see that there is a vase on it. I do not 'see' patches of light or only an array of objects. My knowledge is enriched by (among other things) the proposition asserted by a sentence such as 'The vase is on the table', even though I did not utter (audibly or otherwise) this or any other sentence. As Hanson (1958) argued 'Knowledge of the world is not a *montage* of sticks, stones, color patches and noises, but a system of propositions (p. 26)'. When we use the word 'see', we refer to a bridge between a pattern of sensory stimulation and knowledge which is propositional. This is not to deny that there are such things as appearances, only that if they have a role to play in cognition, the nature of such a role is at present a complete mystery. We cannot even talk about appearances without, in fact, talking about the propositional content of the appearances. And as Wittgenstein wisely reminded us, in such circumstances it is best to remain silent on the subject.

Failure to grasp the difference between the appearance of visual images and knowledge leads to various logical confusions. For instance, it leads Bugelski to conclude that the only reason deaf children should recall visual patterns is because they used visual images which, judging by his use of the term, means that they could recreate the appearance of the pattern by reactivating something like the original sensory stimulation. By such an account each time the image is recalled, it must be 'seen' anew since deaf children presumably do not possess the appropriate code (i.e., an auditory image of words) with which to represent the proposition as a subvocal sentence. If this were so, deaf children (who, incidentally, very likely have a language of some kind) would be left wallowing in appearances without a single item of knowledge to which the asciption 'true' or 'false' was even applicable!

There are even more serious difficulties involved in drawing the distinction between appearances and propositions if we carry this dichotomy through to the area of thinking. The role of experienced images (i.e., appearances) in thinking is by no means clear since even if we make the assumption that the contents of our experiences reveal theoretically useful psychological processes, it still remains true that very little (if any) of the thinking is carried by such processes. Thus, as Humphrey (1951) points out, while the process of thinking ' . . . may involve sense-resembling processes of a particular modality . . . this is the the cart, not the horse. The primary "work" when one thinks a proposition such as "Russia is East of Britain" is imageless ... (p. 106)'. Natsoulas (1970) concurs with this view, arguing that even though thinking can involve a succession of imaginal awarenesses, 'In undergoing the image, however, one does not have the thought. One has it in noticing something about what is imaginally presented. Such noticings are, of course, propositional ... (p. 99)'.

The same could also be said about the role of imaging in other cognitive tasks, including the learning of paired associates. For suppose one maintains that a subject learns a pair such as *boy-play* by forming an image of a small boy throwing a ball. Later when the stimulus word *boy* is presented, the same image is retrieved. By examining this image so it is argued, the subject is able to produce the correct response *play*. The problems remains, however, to explain why the subject in this case chooses to respond *play* and not *throw* or *ball* or *catch* or any of an unlimited number of words equally appropriate to that image. Presumably it is because he remembers more than is contained in the image. In fact, this shows that the bulk of the work of learning and recalling the pair of words is carried out by a process to which we do not have conscious access, but which may in some unspecified manner, make use of the propositional information that *boy* and *play* are related by predication.

Before leaving this discussion of knowledge as propositional, we must pause to reemphasize the difference among pictures, sentences, and propositions. Both pictures and sentences must be interpreted before they become conceptual contents. This is because there are an indefinite number of both pictures and of sentences which are cognitively equivalent. This is not true of propositions as logicians use this term. For example, the philosopher Frege (1960) in his seminal work on predicate logic (first published in 1879) cites the example of two sentences which are paraphrases of one another and comments,

... I call the part of the content that is the same in both the *conceptual content. Only this* has significance for our symbolic language; we need therefore make no distinction between propositions that have the same conceptual content (p. 3).

Thus propositions are to be found in the deep structure of language and not in its surface form. But this is still not sufficiently abstract for our purposes since it might be taken to imply that each proposition is expressable by *some* sentence in a natural language. This is not, however, a necessary condition for our use of the term. We claim that it is still useful to think of propositional knowledge even when the concepts and predicates in such prospositions do not correspond to available words in our vocabulary. Such concepts and predicates may be perceptually well defined without having any explicit natural language label. Thus we may have a concept corresponding to the equivalence class of certain sounds or visual patterns without an explicit verbal label for it. Such a view implies that we can have mental concepts or ways of abstracting from our sense data which are beyond the reach of our current stock of words, but for which we *could* develop a vocabulary if communicating such concepts became important (e.g., for a professional musician, painter, or wine taster). This is not an unreasonable position to hold in view of the fact that conceptual categories are necessary not only for communicating but also for *acting* on the environment. Thus perceptual or motor events which are functionally equivalent with respect to indicating or leading to functionally similar changes in the organism's environment might become represented as unique nonverbal mental concepts (for a discussion of such an action-oriented view, see Arbib, 1972). Such a view is in agreement with Newell and Simon's (1972) position that postulating a single set of internal symbol structures provides the most parsimonious account for both thought and the deep structure of language. It also receives support from evidence (e.g., Macnamara, 1972) that children develop conceptual or semantic structures prior to learning the related linguistic signs.

In spite of the inexpressibility (for a particular individual at a particular time) of propositions containing such nonverbal concepts, there are nevertheless some good reasons for still referring to such knowledge as propositional or descriptive. Just as cognition requires propositions which stand in a type-token relation to sentences, so also does it require something which stands in a type-token relation to pictures or sensory patterns. This something is best characterized as a descriptive symbol structure containing perceptual concepts and relations, but having the abstract qualities of propositions rather than the particular qualities of pictorial images. Furthermore, to refer to a representation arising from sensory stimulation as being propositional, as we have been advocating, is to imply (a) that it does *not* correspond to a raw sensory pattern but, rather, is already highly abstracted and *interpreted*, (b) that it is not different in principle from the kind of knowledge asserted by a sentence, or potentially assertable by some sentence, (c) that it depends on the classification of sensory events into a finite set of concepts and relations, so that what we know about some event or object is formally equivalent to (i.e., can be reduced to) a finite (and, in fact, relatively small) number of logically independent descriptive propositions. The above implications, as we shall see in the next section, are desirable and yet difficult (if not impossible) to convey using the picture vocabulary of the imagery literature.

1.3. Picture Metaphor

In this section we try to make explicit some of the implications of using the imagery vocubulary. To begin, consider what the terms 'image' or 'imagery' mean to most psychologists who write on the subject. Some writers have suggested that images are related to conditioned sensations (Staats, 1968), to "indirect reactivations of former sensory or perceptual activity (Bugelski, 1970, p. 1002)," or that they are a "faint subjective representation of a sensation or perception without an adequate sensory input (Holt, 1964, p. 255)," or "the occurrence of perceptual processes in the absence of stimulation which normally gives rise to perception (Hebb, 1966, p. 41)," or imagery is defined as "the ability of a subject to generate or synthesize a sensory-like datum in the absence of physical stimulation (Weber and Bach, 1969, p. 199!)."

Such definitions, however, are not used directly in the empirical research. As Paivio (1969) rightly points out, "Both images and verbal processes are operationally defined and the concern is with their functional significance . . . (p. 243)." However, the importance of the informal notions of imagery in psychological theories should not be underestimated. What makes it possible to give a consistent and systematic interpretation of the empirical findings is not the individual predictions (e.g., high-imagery sentences are recalled more easily than low-imagery sentences) nor the operationally defined terms (imagery as the rating assigned to a stimulus), but the highly persuasive intuitive notions of what images are, what causal effects they may exert, and how we can manipulate them in our mind.

This can be seen clearly if we consider that various different experimental paradigms require different operational definitions of the construct *image*. Thus in research in which the effects of various nmemonic strategies are compared (e.g., instructions to use images as opposed to other methods), there is one definition of image (image$_1$). In experiments investigating the influence of different stimulus attributes (e.g., high- versus low-imagery words) there is another operational definition (image$_2$). Other research procedures involve the adoption of still other definitions of the theoretical construct *image*. The identity of these various constructs (image$_1$ = image$_2$

= ...) does not, however, follow from any of the operational definitions nor from the results of the experiments (although, of course, similar patterns among empirical correlates of the various manipulations of imagery gives one some grounds for believing that they are related).

The unity of these constructs, and consequently the coherence of the notion of imagery rests on a metatheoretical assumption. This assumption, in turn, rests on the persuasiveness of subjective experience and on the ordinary informal meaning of the word *image*. In this context the term relies heavily on a picture metaphor. The whole vocabulary of imagery uses a language appropriate for describing pictures and the process of perceiving pictures. We speak of clarity and vividness of images, of scanning images, of seeing new patterns in images, and of naming objects or properties depicted in images.

There is, of course, nothing wrong with using metaphors: Virtually all theoretical ideas in science derive from some relatively familiar metaphor. However, not all metaphors are equally appropriate and some may even be harmful by discouraging certain kinds of fundamental issues being raised and by carrying too many misleading implications.

For example, one misleading implication involved in using the imagery vocabulary is that what we retrieve from memory when we image, like what we receive from our sensory systems, is some sort of undifferentiated (or at least not fully interpreted) signal or pattern, a major part of which (although perhaps not all) is simultaneously available. This pattern is subsequently scanned *perceptually* in order to obtain meaningful information regarding the presence of objects, attributes, relations, etc. This "image retrieval before perception" view is phenomenally very powerful and is implicit in the everyday sense of the word "image." It is also present in all the illustrative examples used by psychologists to persuade their colleagues of the reality of images. For example, in discussing the use of the "one-bun" rhyming mnemonic used by his subjects, Bugelski (1968) states,

> The most convincing evidence regarding imagery comes from the reports of many Ss who expressed the belief that they did not know some or any of the words when either the original learning or recall test began. They would then mumble the numeral, state the rhyme word, and then report "oh, yes, hen-ski." They asserted that the "little hen on skis" had to appear before they could report "ski" (p. 332).

Atwood (1971) is quite right when he states, "The most elementary question which can be asked about mnemonic visualization is the following: does the mnemonic image actually involve the visual system (p. 291)?" Using a method of selective interference, he gathers evidence which leads him to conclude that to a large extent it does. He writes,

> Verbal material may be recoded into a visual image (e.g., during application of a mnemonic device) and encoded into memory as a primarily visual schema. During recall, the schema is *decoded visually* and then recoded once again into verbal symbols (p. 297, italics added).

Similarly Bahrick and Boucher (1968) argue in favor of an "imagy retrieval before perception" view. They write,

> if one is asked to recall the color of a couch in the living room of a friend's home, however, it is likely that the verbal transformation occurs at the time of recall, and is based upon stored visual information (p. 417).

This is exactly the same type of argument which Shepard (1966) used,

> ...if I am now asked about the number of windows in my house, I find that I must *picture* the house, as viewed from different sides or from within different rooms, and then count the windows presented in these various mental images (p. 203).

The view depicted in the above quotations, though phenomenally quite sound, presents serious problems if it is taken as an explanatory account of the process of retrieving pictorial information (i.e., information initially acquired visually) from memory. This is because however metaphorically one interprets the notion of picturing a recalled scene in one's mind, the implication is always that whatever is retrieved must be perceptually interpreted (or reperceived) before it becomes meaningful. In other words, the appearance of a memory image precedes its interpretation by the usual perceptual processes, such as those resulting in figure-ground distinctions, object individuation and identification, and the abstraction of attributes and relationships among elements of the scene. But what can serve as the input to such a perceptual process? Whatever it is, it must be very much like the pattern of sensory activity which takes place at various levels of the nervous system when some sensory event token occurs.

Such a position, however, runs into many difficulties. First, in supposing that information received through the senses is stored in memory and retrieved at a later date in an uninterpreted form, we place an incredible burden on the storage capacity of the brain. In fact, since there is no limit to the variety of sensory patterns which are possible (since no two sensory events are objectively indentical), it would require an unlimited storage capacity.[1] Second, such a view creates severe difficulties for the retrieval process. Since the sensory events are stored in "raw" form, retrieval can occur by one of two means. Either one retrieves an image by some sort of scanning process in which putative candidates are placed before the mind's eye to determine whether they are appropriate, or else the images are tagged by some gross labels and associatively retrieved by a multiple-sort key. The first of these is unreasonable on several grounds. Perceptual processes in the "mind's eye" appear to be no faster than the usual perceptual recognition processes, so the time for an exhaustive search would be prohibitive. Furthermore, the conscious awareness, which suggested image storage in the first place, also reveals that we directly retrieve the correct information without a series of false attempts.

The second alternative is implausible on the grounds that we can retrieve information about a whole scene or any part of it by addressing aspects of the *perceptually interpreted content* of the scene. Even if we confine ourselves to the retrieval of phenomenal images, we can argue that the content of such images must be already interpreted — in spite of the fact that we seem to be "perceiving" them as we would novel stimuli. This must be so because retrieval of such images is clearly hierarchical to an unlimited degree of detail and in the widest range of aspects. Thus, for example, I might image a certain sequence of events at a party as I recall what happened at a certain time. Such images may be quite global and could involve a whole scene in a room over a period of time. But I might also image someone's facial expression or the jewel in their ring or the aroma of some particular item of food without first calling up the entire scene. Such perceptual attributes must therefore be available as interpreted integral units in my representation of the whole scene. Not only can such recollections be of fine detail but they can also be of rather abstract qualities, such as whether some people were angry. Furthermore, when there are parts missing from one's recollections, these are never arbitrary pieces of a visual scene. We do not, for example, recall a scene with some arbitrary segment missing like a torn photograph. What is missing is invariably some integral perceptual attribute or relation, for example, colors, patterns, events, or spatial relations (we might, for example, recall who was at the party without recalling exactly where they were standing). When our recollections are vague, it is always in the sense that certain perceptual qualities or attributes are absent or uncertain — not that there are geometrically definable pieces of a picture missing. All of the above suggest that one's representation of a scene must contain already differentiated and interpreted perceptual aspects. In other words, the representation is far from being raw and, so to speak, in need of "perceptual" interpretation. The argument is not simply that retrieval of images would involve a bewildering cross-classification system while retrieval in other forms of representation would not. The point is that because retrieval must be able to address perceptually interpreted content, the network of cross-classified relations must have interpreted objects (i.e., concepts) at its nodes. Thus storing images at these nodes as well is functionally redundant. This does not mean, of course, that what we retrieve cannot be further processed. We shall examine several ways in which such representations can reasonably be thought of as being processed further after retrieval (e.g., by the application of operations such as counting). The argument is simply that they are not subject to *perceptual* interpretation the way pictures are interpreted.

Our attack against the notion of an image being an entity to be perceived need not, of course, appeal to phenomenal observations. Consider, for example, the following argument. There are denumerably many logically independent propositions true of any scene or of any physical object (including a real picture). Since the brain can store only a finite (in fact, relatively small) amount of information about any one scene, we might ask about the nature of this selected finite subset. One possible answer is that the stored representation is a pictorial image of limited resolution (i.e., one which can effectively be replaced by a finite two-dimensional grid, each element of which contains a selection from a finite set of attributes). But this is unsatisfactory not only because it still leaves too much information (we can easily show, because of the fineness of some of the details recalled, that the overall resolution of any pictorial representation would still have to be rather high), but because such an approach results in the wrong *kind* of information being selected. Thus, as we argued in the previous paragraph, we are more likely to recall such things as which objects were present without recalling their exact relations than we are to recall all the detailed information but with low precision.

We may assume, then, that the representation differs from *any conceivable* picture-like entity at least by virtue of containing only as much information as can be described by a finite number of propositions. Furthermore, this reduction is not reasonably accounted for by a simple physical reduction such as that of limited resolution. What type of representation meets such requirements? A number of alternative forms of representations are discussed in a subsequent section. For the present, it suffices to point out that any representation having the properties mentioned above *is much closer to being a description of the scene than a picture of it*. A description is propositional, it contains a finite amount of information, it may contain abstract as well as concrete aspects and, especially relevant to the present discussion, it contains terms (symbols for objects, attributes, and relations) which are the *results* of — not inputs to — perceptual processes. Of course, to say that the representation was a description without being more specific about the nature of such descriptions still leaves it vulnerable to some of the types of criticisms which we have directed against images. Both images and descriptions carry too much undesirable excess meaning; for example, the latter may imply a fixed-order of access, as in reading, which is certainly unwarranted. "Descriptions" of the type we have in mind are never accessed in a fixed serial order in any of the systems which we will examine later. Apart from the arguments made above and those which we mentioned earlier in discussing knowledge as being propositional, the notion of a description gains its greatest advantage from the fact that it has been formalized in a number of areas (e.g., in computer-simulation models). In such contexts the representations provide a formally adequate amount of certain types of cognitive activity while, at the same time, corresponding closely to what we intuitively mean by the term "description."

The mental representation differs from what is inferred from the conscious image in many ways. For example, to use an illustration cited earlier, while two visual images of a chessboard may be pictorially identical, the mental representation of one might contain the relation between two chess pieces which could be described by the phrase "being attacked by" while the representation underlying the second image might not (cf. Simon & Barenfeld, 1969). For this reason, it would be reasonable to expect that the mental representation of a configuration of pieces on a chessboard would be much richer and highly structured for a chess master than for an inexperienced chess player. This view is supported by de Groot (1966) who found that chess masters could recall an authentic board position much better than inexperienced players (after viewing it for 5—10 seconds) in spite of the fact that their *visual* memories were no better (as measured, say, by their ability to recall chessmen randomly placed on a board).

As another example, it would be quite permissible, according to the view which we have been presenting, to have a mental representation of two objects with a relationship between them such as "beside." Such a representation need not contain a more specific spatial-relation term such as "to the left of" or "to the right of." It would seem to be an unreasonable use of the word "image," however, to speak of an image of two objects side by side without one of the relations between them being either "to the left of" or "to the right of." (The fact that children, who are especially adept at "visual imagery," frequently have difficulty in discriminating a figure such as a letter from its mirror image, suggests that their mental representation of such figures suffers precisely from such a lack of explicit differentiation of the relations "to the left of" or "to the right" in favor of a more general relation such as "adjacent to" or "away from the center.") Similarly, we could have a mental representation of a triangle which might consist of a structure in which the symbol "triangle" was hierarchically linked (by the relation "has as parts") to three representations of lines which were, in turn, linked to each other via relations labeled "connected to." Such a network (which resembles many of the artificial intelligence data structures – see below) need not contain relations of the type "at an angle of n degrees to." On the other hand, there is considerable uncertainty (as dramatized by the debate between Locke and Berkeley) regarding the possibility of having an image corresponding to the above representation, namely equilateral, equicrural, nor scalenon; but all and none of these at once."

To summarize, then, we have argued that the functional mental representation is not to be identified with the input to a perceptual stage but rather with the output of such a stage, inasmuch as it must already contain, in some explicit manner, those cognitive products which perception normally provides. If we could think of functional (rather than phenomenal) "images" in this sense, we would have removed the disturbing duality of "image" and "mind's eye," while, at the same time, we would have answered some of the puzzling classical questions referred to earlier: An "image" *qua* representation in our sense can certainly be selective, generic, abstract, and even unconscious inasmuch as the cognitive products of perception can be all of these.

2. ALTERNATIVE FORMS OF REPRESENTATION

In this section we briefly examine three approaches which have been used in theoretical studies of the representation problem for cognition. The approaches are closely related and are distinguished primarily by the research areas in which they are developed and by the descriptive formalisms which they employ, although there are one or two more significant differences among them which we shall try to draw out. The first approach involves the use of propositions and usually relies on deductive proof procedures for processing them. The second approach derives primarily from work in computer simulation of cognition and in artificial intelligence. The form of the representation is called an information or data structure and is frequently described in terms of directed graphs. The third approach represents concepts in terms of procedures. These three types of approaches are described below.

2.1. Propositional Representations

Because of the availability of the predicate calculus as a formal language for expressing the contents of knowledge, propositions have been widely used — especially by students of artificial intelligence — as an explicit form of representation. This form has the great advantage that well-known mathematical systems for manipulating formal sentences can be applied to the representations to derive their logical entailments.

In its simplest form, such an approach assumes that what a person knows can be represented by a finite list of propositions or axioms (although, to repeat again an earlier point, this must not be taken to imply that tokens of actual sentences in some natural language are stored). Rules of deductive reasoning can then be applied to this list to generate all the logically valid propositions which follow from the initial "premises." Herein lies one of the attractions of this approach: It is generative in the sense that an unlimited number of "beliefs" can be deduced by a straightforward mechanical procedure from the initial representation. Thus it ought to allow an indefinite number of questions to be answered about the knowledge represented.

Question-answering systems, in fact, have been developed which represent their data base in the predicate calculus and which use a theorem-proving procedure for retrieving information (e.g., Green and Raphael, 1968). The question to be answered is converted to a proposition to be proven in the system. A constructive proof of this proposition then provides the answer. For example, a constructive proof of that there exists an object (in the formal sense of this term, including mathematical objects such as numbers) which satisfies certain conditions would actually identify such an object. Slagle's question-answering system DEDUCOM and the more comprehensive MULTIPLE (Slagle, 1971), which can be applied to a wide spectrum of problems from playing chess to solving problems in logic, both use an explicit propositional data representation. The Stanford Research Institute robot "Shakey" (Raphael, 1968) operates by storing its "knowledge of the world" in propositional form and using a theorem prover to respond to commands. For example, the command to push two large blocks together is transformed into a proposition to the effect that there exists a path of travel which leads to the desired state with two large blocks together, given the present conditions and known constraints. A constructive proof of this proposition would derive a path satisfying the requirements, which would then be converted into a sequence of overt motions by the robot.

In spite of considerable success with this form of representation of knowledge in a variety of artificial intelligence applications, it does have some serious limitations. In addition to the general problems associated with the use of the predicate calculus to represent knowledge of a changing environment (e.g., the "frame problem" discussed by Raphael, 1971), there are additional problems which appear when we think of such a system as a model of human cognition. In order for such a system to have "psychological reality" it must take cognizance of empirical data concerning the psychological complexity of various cognitive tasks (for a discussion of this point, see Pylyshyn, 1973). In other words, it must account for empirical data such as that made available by chronometric analyses of a variety of recall, verification, and the problem-solving experiments. Theorem-proving processes, as well as the specific propositions posited to constitute the representation of knowledge, must reflect the relative complexity of various cognitive tasks as inferred from empirical studies. In addition, the system must display similar *intermediate states of knowledge* as subjects do in solving a problem or answering a question. For example, part way through attempting to answer a question, a subject may have the answer to some other related questions. An adequate model should account, in a general way, for such sequences of partial solutions. It is not clear at this stage in our understanding of theorem-proving schemes whether a uniform proof procedure is capable of meeting such requirements. If it were to do so, however, it would have to be molded to fit empirical data in at least two ways: (a) by the selection of starting propositions (which may include derivable theorems as well as independent axioms) and (b) by the selection of an appropriate proof method. These are discussed below.

Consider that there are an indefinite number of sets of base propositions which can serve as logically equivalent representations (i.e., from which the same ultimate set of propositions can be derived). From the standpoint of a logician, the smallest number of simple logically independent axioms would be preferred. From the standpoint of a psychologist interested in describing a mental representation of knowledge in a certain domain, this is only one criterion. He is interested in the simplest representation which accounts not only for what is known, but also for empirical evidence concerning such properties as accessibility. Thus while it is logically immaterial which of two propositions, "A is larger than B" or "B is smaller than A," is contained in the representation, the two are not equivalent from a psychological point of view. Which proposition is expressed in a problem description affects how difficult the problem is to solve (Clark, 1969). One could point to the fact that the predicate "is smaller than" is *marked*, that is, it has both a nominal and a contrastive sense. From this one could argue that the sentence "B is smaller than A" may be psychologically represented by three propositions, such as "A is larger than B," "B is small," and "A is small" (Clark would represent it as "A is small· B is small+" where "small+" signifies "smaller to a greater degree"). Indeed, such a hypothesis appears to fit the available empirical evidence (Clark, 1969).

Another potential source of development may come from studies in computational complexity as applied to theorem-proving systems. For example, as it stands, one of the difficulties of a logic-based model of cognition is that if any pair of propositions in the representation is contradictory, the whole system breaks down (since anything can be proven in a system in which both *p* and *not-p* are axioms). If we had a measure of derivational complexity — or a measure of the distance of a derivational path between two propositions — which was based on psychological considerations, the problem of contradictions could be dealt with. In this case, it might be reasonable to tolerate contradictory propositions so long as the derivational path between them exceeded a certain minimum value. Such a proposal was, in fact, made recently (Arbib, 1969). If we distinguish, as do Simon and Newell (1956), between logical entailments (all statements derivable from the axioms) and psychological entailments (all statements which are evidently true, to a person, as a consequence of the axioms), then we have the basis for an interesting extension of a predicate calculus model of cognition. We would identify the psychological entailments as those statements derivable from the base set by a path of less than a certain critical length.[2] Such a system would have the interesting consequence that it would allow a person to hold contradictory beliefs and to make contradictory statements without being aware that they were contradictory!

2.2. Data-Structure Representations

The idea of general data or symbol structures grew out of several pioneering achievements in computer science. One was the work of the group at the

Carnegie Institute of Technology beginning in the late fifties which led to the first list-processing system known as the Information Processing Language (see the historical notes in Newell and Simon, 1972). Another was the work on computer graphics which was pioneered at the Lincoln Laboratory at Massachusetts Institute of Technology (Roberts, 1965; Sutherland, 1963). Both of these were attempts to represent information in a manner best suited to the processes which would operate on it. The design of *appropriate* data structures is one of the central tasks of computer science, and many difficult problems, such as that of processing graphical data for display on an oscilloscope screen, were solved only after clever new forms of representation were developed. An appropriate representation for a particular information-processing application is one which (a) contains symbols which designate the functionally important and most invariant *aspects of the environment which is being represented* and (b) gives the processes access to a variety of units of data, from individual primitive symbols through overlapping subsets of related symbols up to the entire representation. There must, in other words, be a facility whereby symbols can designate symbol structures in which individual symbols can designate still other symbol structures, etc., in both a hierarchical and heterarchical fashion.

A wide variety of data structures have been developed for different purposes. They are usually depicted as directed graphs in which nodes represent symbols (which may, in turn, designate other symbol structures of objects in the environment or even programs) and links represent relations of various types (i.e., the links may be labeled according to the type of relationship they represent, e.g., "is connected to," "is a part of," "is an instance of," or "has the property"). Because such representations are extremely varied in form as well as in the way they function in different systems, and because they are rather common in the information-processing literature, they are not discussed in detail in this paper. For further elaboration, the reader is invited to consult the historical papers of Sutherland (1963), Roberts (1965), or the papers contained in Minsky (1968) where a variety of data structures are discussed. Reitman (1965), Newell and Simon (1972), and Frijda (1972) also present a discussion of the use of simple list structures in psychological theories. Such data structures meet many of the basic requirements for a cognitive representation: Only functionally relevant aspects of the environment are mapped onto the representation, distinct representations mean functionally distinct stimulus *types*, and relations among stimulus types can be accounted for by relations among representations (i.e., by the presence or absence of nodes or links in the underlying data structure). In fact, the contents of data-structure representations may be viewed as *propositional*. By identifying links with predicates and nodes with designating expressions (or in some cases with other propositions), we can generate a finite set of propositions (e.g., "line X is part of figure A") which exhaustively describes the knowledge which the system has of the environment.

In spite of the close relation between data structures and propositions, there are a number of important differences between them. A list of propositions has little inherent structure. While certain relations among the propositions may be implicit in the way in which various symbols occur in them or the way in which the propositions tend to be used in groups to prove theorems, this structure is of a rather limited kind. Relations among terms are much more explicit in data-structure representations because of the explicit access relations provided by the system of links. This usually makes the data-structure network more useful and natural for artificial intelligence applications.

2.3. Procedural Representations

The third form of representation which we shall examine is one in which concepts and facts are represented in terms of rules or procedures. The view that what is stored is a system of rules or a procedure is an attractive one on many grounds and has enjoyed popularity in a number of circles (see, e.g., Davies and Isard, 1972; Miller, Galanter, and Pribram, 1960; Pylyshyn, 1973; Winograd, 1972). An obvious argument in favor of a rule description is on the grounds of descriptive economy: a small number of rules can cover a wide domain of instances. Another argument is the intuitive idea that what we know when we have learned something (say a concept) is how to use it. This is related to the notion of operational meaning and to the position (made famous in the 1930s by Rudolf Carnap) that the meaning of a word is bound to the method by which statements containing the word are checked for truth or falsity.

Intuitively, it seems clear that at least part of what we know when we have learned a concept includes a set of specific procedures for determining whether a particular token is an instance of the concept as well as a set illustrates how the program makes use of a variety of specific situations. In other words, we not only know facts but also how to take certain actions relevant to the facts. From such considerations it is possible to argue that the representation of certain concepts is nothing more than the set of such procedures. We shall take the position that while this claim is undoubtedly true, it may also be somewhat misleading in its usual interpretation. We shall return to this point in the latter part of this section.

One of the earliest proposals for including procedural predicates in a propositional system was made by McCarthy (1959) and has been the source of several subsequent developments. The most successful recent attempt to exploit the notion of procedural representation is a system for understanding natural language developed by Winograd (1972). Winograd's system is a computer program which maintains a sophisticated model of the knowledge which a robot needs to operate in a limited environment. The robot is assumed to be equipped with an eye and a hand. Its simulated environment consists of a collection of blocks of various sizes and colors which it can manipulate. The system can enter into a dialogue with a person concerning this environment. It can understand declarative English sentences about the environment and add the information conveyed to what it already knows. It can interpret and simulate the execution of commands related to manipulating objects in the scene (i.e., it can change the representation of the location of objects: There are no actually physical objects, and the machine does not have a real perceptual motor device). It can also answer a wide range of English questions both about the scene and about its own actions. While the most impressive aspect of this system is the way in which the various subsystems work together to produce intelligent behavior, our concern here is only with the question of how the system represents its knowledge. This knowledge includes not only knowledge of objects in the scene, but also knowledge of grammar, semantics, and deductive logic. For simplicity we will concern ourselves with only one aspect of the total system — that which illustrates how the program makes use of procedural information in its representation.

The form of representation adopted by Winograd contains aspects of both the data structure and the propositional forms of representation discussed earlier. Recall that one of the defining characteristics of data structures is the presence of explicit access links which enables the tracing of paths through the structure in a straightforward data-governed manner. In contrast, one of the defining characteristics of the propositional representation is that inquiries to it are dealt with by a neutral and uniform proof procedure whose operation does not depend on either the inquiry or the data. Each of these two approaches has its advantages. In the data-structure case, by making as many as possible of the relations among concepts and among substructures explicit in each representation, we gain considerable access efficiency. It is no longer necessary to refer to the entire data base and to perform complex computations to go from one substructure to another, since much of this has been done for us in advance. In the propositional representation, going from one set of propositions to another is done by a single uniform proof algorithm which, being independent of the data, must consider the entire data base as being relevant to all deductions.[3] On the other hand, the uniform proof method allows us to get at a wide range of questions, including ones not initially anticipated, without making changes in the way the data is represented. This gives the propositional representation an important advantage over data structures.

Winograd proposed an alternative representation combining the efficiency of the "relationship-as-part-of-the-representation" characteristic of data structures with the generality and descriptive-uniformity characteristic of deductive propositional systems. This is done by adopting a theorem-proving deductive system in which the procedures to be used are not neutral with respect to the data to which they are applied, but rather, the data-base representation contains directions as to how to go about proving assertions about particular concepts in the data base. In effect, the propositional knowledge contained in the representation is expressed in an imperative rather than in a declarative language. As an example, take the proposition which might be expressed in English as something like "An object is an X (e.g., chair, sentence, thesis) if it has property x or property y but not property z." Instead of this assertion, we would have a hierarchical, goal-oriented, and partially ordered sequence of procedures which might be interpreted something like,

If you wish to show that an object is an X, then check first whether it has property z: Do this by trying the following procedures ... or, if they fail, by trying the following. ... If any of these succeed return a FAIL. If they fail try next to show that the object has property x or y by referrring to all assertions mentioning these properties or all procedures having these properties as a consequent. ...

Such an imperative or procedural representation is able to state a specific order in which to try out tests, to recommend heuristic short-cut procedures, to specify procedures in terms of goals, and to suggest sections of the data base in which to search. Furthermore, new procedures do not have to be attached to a particular place in the representation (i.e., linked to a particular concept) but may simply be added without relating them directly to the rest of the data base. The general instruction to "try proving that ..." will locate relevant procedures. The way they are stated makes it possible to see when procedures are relevant. Such an approach is made possible by the availability of a general goal-directed imperative language called PLANNER (Hewitt, 1971). The resulting system provides a powerful and efficient representation of knowledge capable of accommodating both facts and data-dependent ways of relating facts. It is, however, essentially heuristic in nature, that is, it depends primarily on logically incomplete short-cut methods. While, if all else fails, it *could* be made to resort to a uniform proof method, this is considered to be unnecessary. The basic procedures are designed to be "sensible" methods of going about relating things and may even suggest at some point that the goal is likely to fail so the system should give up looking any further.

Such a procedural representation is an extremely attractive idea from many points of view, both as an approach to constructing artificial intelligence devices and as an approach to the problem of cognitive representation. Elsewhere (Pylyshyn, 1972, 1973) the author has argued that one has to be particularly careful in selecting the procedures which are to define the representation. One can get into difficulties by taking the most obvious heuristic procedures such as one might infer, for example, from an analysis of think-out-loud protocols. While it is beyond the scope of this paper to present these arguments here, it might be appropriate to indicate briefly what these difficulties are.

As was argued in connection with our earlier discussion of imagery, it is unlikely that processes of which we are aware will turn out to be useful in an explanatory theory. We have already pointed out several ways in which criteria of adequacy for an explanatory account are rather different from those which might be appropriate for an informal descriptive account. If it is to serve as an explanatory account of what a person knows when he has mastered a certain concept, the representation of that concept very likely has to contain procedures more abstract and general than those moment-by-moment procedures which a subject is aware of using. This comes about because the theorist's task in accounting for how a certain concept is represented involves more than simply describing the procedures which a person might use to assign an instance to that concept *in certain typical situations*. The fact that as novel instances are presented to him, a subject can keep coming up with clever new heuristic procedures for assigning those instances to concepts (e.g., for deciding whether strings of words are sentences), and may even resort to external mnemonic aids as the task becomes difficult, suggests that his representation of the concept is not limited to a finite list of such consciously available procedures. Rather, he is able to creatively generate new heuristic procedures from a representation which, while it is most likely procedural, is itself more abstract than a list of the procedures he is aware of using on specific occasions. The underlying abstraction characterizes what Chomsky (1965) has called the subject's *competence* and is discussed in some detail in Pylyshyn (1973). While it is procedural, a competence characterization is not heuristic. It attempts to be complete, that is, to describe the mental representation in a manner which accounts for all the cognitive distinctions in a certain theoretical domain which could be made considering all conceivable circumstances.

3. INFORMATION PROCESSES AND IMAGERY

The discussion so far has been concerned primarily with the question of how knowledge might be represented in memory. Let us now consider the issue of how we might characterize the representations and the processes which are involved when a person is engaged in what he calls imaging. Before we can proceed with these rather speculative suggestions, we must introduce some general remarks regarding differences in levels of knowledge or in types of representation which enter into various stages of cognition. This, in turn, leads us to consider different levels of accessibility or of activation of these representations.

In his excellent analysis of the nature of complex systems, Simon (1969) makes it clear that there are powerful reasons, both from the point of view of the evolution and operation of a system and from the point of view of the scientist's ability to understand the system, for it to be organized in a hierarchical fashion. Such reasons can be used to argue that in studying cognition we ought to distinguish levels of knowledge. What a person knows would then be described as being hierarchically organized. Levels of this hierarchy might be distinguished, for example, on the basis of universality or permanence (in relation to external modifiability). Thus we might distinguish among universal and innate properties of cognition, properties which develop gradually with maturation and general experience, properties having to do with particular domains of knowledge (including domain-specific operational knowledge concerning how to deal with certain concepts), and properties having to do with particular instances, that is, representations arising from particular events, or novel constructions generated in the course of solving a particular problems or in generating some particular overt behavior. It is reasonable to expect that these levels of knowledge may have to be treated in a somewhat different manner within a cognitive theory. One of the virtues of a theory such as Winograd's (1972) system for understanding natural language that it incorporates a distinction among levels of knowledge very much like the one outlined above.

Another general consideration, also related to Simon's (1969) arguments for hierarchical organization, which suggests that one might usefully treat some classes of representation in a somewhat distinct manner, has to do with questions of efficiency of access. Efficiency may be gained through the use of levels of activation or accessibility, with a few items being highly accessible and larger numbers being progressively less accessible.

As an example of the notion of hierarchical accessibility, consider the following: Suppose a process (computer or human) makes use of a certain repertoire of n items of information, numbered from 1 to n. Suppose further that the currently active subprocess makes repeated reference to Items 2, 3, 5, and 7, and that at the present moment an operation is being performed on Item 3. In such a situation there is considerable virtue in arranging for the sets $\{1, 2, 3, \ldots, n\}$, $\{2, 3, 5, 7\}$, and $\{3\}$ to be differentially available or to be at different levels of preparedness or activation. Thus, for example, in some computers Item 3 might be placed in the accumulator or other special register; Items 2, 3, 5, 7 might be placed in some designated common-communication area while the remaining items would remain in general memory. Of course, items need not in any sense be moved about; they might simply be placed in some more ready state (e.g., in the computer example, the addresses of the items might be listed in some stack). We might then identify such an active or ready state with a cognitive buffer or a *workspace.*

Such a workspace would have several additional values. It would provide a stage at which items closely related to a particular item being processed could be held in readiness. This corresponds to the well-known psychological phenomenon sometimes referred to as reintegration, wherein retrieval of part of a structure of related items (e.g., recall of one word of a sentence) results in the recall of the whole structure. It would also provide a stage at which a representation being recalled could be restructured into a form more appropriate for a particular task at hand (more appropriate, that is, than the form in which it was originally stored).

This workspace would also be useful as a stage at which general computational processes are applied to representations. For example, consider what would have to happen when the concept of *number* and that of *window* and of *my house* are being related to one another to answer a question regarding the number of windows in my house. It would be unreasonable to hold that "number of windows in my house" is a static term in my store of explicit knowledge. Indeed it *could not* be the case in general since there is no limit to the number of propositions of the type "number of Xs is N" which a person potentially possesses (since there is no limit to the number of designating expressions such as X which he could generate). Answering a question such as the one regarding number of windows, therefore, must depend on the application of a concept of a number to generate a counting procedure which would, in turn, generate the appropriate concept "number of windows in my house." In fact, this is very similar to the way in which Winograd's (1972) system would answer such a question.

The point of this illustration is to suggest that cognition requires the interaction of abstract concepts such as that of *number* with less abstract ones such as *window* and even more particular ones such as *my house*. Considerable computation must occur in the course of such interactions during which the concepts should be in some state of recruitment. It is useful to think of such a stage which several concepts are simultaneously active, as one in which the concepts are held in a buffer or workspace.

It might be remarked that the process of activating a representation or of "placing it in the cognitive workspace" is invariably constructive since most, if not all, concepts are constructive or generative (cf. Niesser, 1967). That is, a complete representation may not simply be placed in a state of alert, but rather a static instance, undoubtedly more specific in its detail than individual stored concepts, may be constructed from such concepts. It would even be reasonable to suppose that a more detailed representation may be generated in the workspace than is, in fact, called for by a particular cognitive task. In this way a savings in number of separate access steps to the main memory may be achieved by retrieving extra information at each access cycle.

Such considerations might suggest that we are tending towards the view (favored, e.g., by Chase and Clark, 1972) that while picture-like entities are not stored in memory, they can be constructed during processing, used for making new interpretations (i.e., propositional representations) and then discarded. This approach views the content of the workspace as a *model* which satisfies the stored propositions. There is little harm in using the metaphor in this context so long as one can resist the temptation of assuming that the relation of the model to its cognitive representation is like the relation of any physical object to its representation. In fact, the possible descriptive interpretations that can be given to a model is a small subset of those which can be given to a physical object. This is because only a small subset of the properties of a model are relevant to its functioning *as a model*. Which particular properties are relevant can only be determined by referring to the description from which the model was constructed. Thus while a physical model or analog has many properties not contained in, or, in fact derivable from, the stored representation (e.g., with a physical model of a molecule, one could determine its weight, color, taste, angular momentum, etc.), these are not used as bases for making inferences from the model. In fact so long as the physical object is bing used *as a model*, all inferences drawn from it were entailed by the propositional representation (plus other stored knowledge) from which it was constructed. Thus the model introduces no new information although it serves the invaluable function of making what was implicit in the description more explicit, accessible, and manipulable. This, of course, is of central importance in cognition. For example, by using heuristics which operated on a diagram, Gelernter's (1963) geometry-theorem-proving system was able to achieve a 200-fold savings in number of search operations. Nevertheless, if we accept the above argument regarding the way in which a model functions, we see that the particular extended physical nature of the model is irrelevant since the model functions like a highly selective abstract and interpreted *percept* – in other words, like a description again.[4] Its importance arises from the fact that it makes possible certain kinds of restructuring and reconstruction of descriptions. But we do not require a picture-like entity to do this. Symbolic descriptions too can be manipulated so as to make various aspects more accessible to certain inquiries. Furthermore, such an approach has the advantage that it does not require positing two qualitatively different entities, one, an abstract propositional-descriptive structure serving for memory representations and the other a picture-like entity with implications of concreteness, spatial extent, and simultaneous availability (all of which must be metaphorically interpreted since none of these terms are intended to apply literally to brain structures) serving in thinking. To make the latter remarks more concrete, we shall devote the next section to describing a system which uses that approach.

3.1. Information-Processing Model

An excellent illustration of the way in which higher and lower level representations might be handled in a cognitive theory and of the use of something like a cognitive work-space is to be found in the recent work by

Baylor (1972).[5] Baylor's system is a cognitive theory designed to characterize (by simulation) the psychological processes involved in solving certain kinds of "block visualization tasks." An example of such a task is the following: "The four narrow sides of a 1-inch x 4-inch x 4-inch block are painted red. The top and bottom are painted blue. The block is then cut into 16 1-inch cubes. How many cubes have both red and blue faces? How many have no painted faces?"

Baylor's work is in the best tradition of information-processing theories and is clearly free of the conceptual difficulties discussed earlier in this paper. Yet, it is addressed directly to the phenomenon of imagery. Because of this, it sheds some light on the question we are currently examining, namely, what is the nature of the information-processing function which accompanies imaging? Consequently, we shall examine his system in some detail.

In a manner somewhat analogous to the "dual-code" schemes in the imagery literature, Baylor distinguishes between what might roughly be described as "factual" knowledge and the more "pictorial" or "imaginal" knowledge. This distinction, as we shall see, is quite different from Bower's distinction between propositions and appearances. Baylor's distinction is made precise in his system. As we examine it in detail, we will find that the difference between the two types of knowledge is not at all a difference in kind but rather a difference in arrangement which results in somewhat different *access relations* in the two cases.

The distinction arises in Baylor's theory through his postulate of two separate but closely related systems in which information about the problem environment is represented. These are called the S space (for symbolic factual information) and the I space (for imaginal information). The idea was to represent in the S space, "information that is true about pieces and their components *in general*; and to store in the I space, information that is true for a *specific* piece and its components (Baylor, 1972; see footnote 8)." In fact, this is not quite accurate since various processes do affix problem-specific information (such as about the color of various faces) to the S-space representation. Such information is, however, represented in a more global manner than it is in the I space as we shall see later.

The S space consists mainly of a data-structure showing how certain atoms are interrelated hierarchically. The atoms are faces, edges, and vertices and are distinguished only as being top, bottom, left, right, front, or back. Thus this structure corresponds rather closely to the statements which we would make about blocks in general *without being able to label or to point to certain particular points on some arbitrary three-dimensional block.* Consequently, the structure does not refer to any *particular* edge or vertex. As a result, many distinct vertices receive the same designation (e.g., TOPVERT) and many particular vertex on a block would be referred to by various designations (e.g., the "top" vertex of the "left" edge of the "front" face would also be the "front" vertex of the "left" edge of the "top" face).

In contrast to this description of a base block, the I-space representation does have attributes and relations which depend on a three-dimensional frame of reference. Thus in I space, vertex atoms such as TOPVERT are assigned attribute values which are symbols for a particular vertex on a block (say, V_1) while edge atoms such as LEFTEDGE are assigned symbols which refer to a particular line on a block (say $V_1 - V_2$). Whenever two atoms refer to the same edge or vertex of a block, these atoms would be assigned the same attribute value. In addition, the *I*-space structure not only displays relations such as "is a part of" (as was the case in the *S* space) but also certain spatial relations such as "is above," "is in front of," or "is to the left of." Thus the *I*-space representation captures more of the structure of the integrated physical object than does the *S*-space representation which is constrained to follow closely the type of verbal description in which block visualization problems are originally stated.

The two representations continue to be distinct but closely related as the problem-solving process continues. As more blocks are created in the *I* space (by slicing the original base block), each is assigned a cross-reference to the *S*-space block, while at the same time a list is kept in the *S* space of the blocks created in the *I* space. Also, if certain faces of a block are painted, the color names are assigned directly to particular faces (say, $V_1 - V_2 - V_3 - V_4$) on the *I*-space block, while in the *S* space the reference to the *I*-space *block* is assigned a list structure description such as "(SIDES COLOR RED) and ((TOP BOTTOM) (COLOR BLUE))."

Thus much of the information is, in effect, stored twice, once as a direct attribute of particular atoms in *I* space and again as a general attribute of the block in the growing structure of "factual" knowledge represented in *S* space. The main difference between these two forms of representation is that certain information is represented directly in the *I* space whereas it would have to be deduced indirectly from the *S*-space representation (perhaps in some cases only with the aid of additional knowledge concerning the properties of three-dimensional objects). This means certain operations, such as counting, can be applied directly to objects in *I* space but not to those in S space. In other words, the main difference between S-space and I-space representations is in the relative accessibility of different aspects of the information to different psychological processes.

In Baylor's system the S space and I space are only roughly hierarchical in our sense of hierarchy of knowledge. Furthermore, much of the higher level knowledge (e.g., the concept of a *slice* or of *number*) is implicit in the permanent operators built into the system. Also, because of the limited domain of application of the system, the range of its higher level knowledge is rather limited. If the system had been a more general theory of cognition, the S space might have included a great deal of information about geometry and about three-dimensional blocks in general, while the I-space representation would still have been the same, since it is adequate to solving block-visualization problems. Of course, other types of tasks might demand a different I-space representation which would then have to be constructed from the S-space information. The main principle of the cognitive workspace, however, does apply to the I-space representation: The data structure in I-space is both more specific than that in the S space and is in a form more appropriate for applying typical operators needed in solving the block visualization task.

It is not difficult to think of phenomenal correlates of many aspects of Baylor's theory. Indeed, the system was designed to bear a close relation to a subject's "thinking-out-loud" protocol produced while he was solving a block-visualization problem. Such a protocol is naturally couched in the language of experience, with persistent references to images and operations on the imagined objects. Thus Baylor's work is proposed as a bridge between certain aspects of the consciously accessible phenomena of cognition (in a restricted domain) and the requirements of an information-processing level of analysis. Such requirements necessitate the development of precise and logically sound (noncontradictory) definitions of constructs. It is of interest, then, to see what happens to the picture metaphor when it is subjected to such demands.

Consider the formal nature of such notions as "image," or "reading an image" in Baylor's system. In his conclusions, Baylor makes the following summary statement:

But what do these various techniques tell us about the use of visual mental imagery in the human thought processes? For one thing, *visual mental imagery is just another representational system*, albeit one that happens to be very convenient for structuring information that was at one time "known" and encoded through the visual system.... Most importantly, perhaps, the processes identified to read the images (in

½–3 s.) are composed of the same kinds of elementary processes identified elsewhere for generating and testing, comparing, counting, and the like (Baylor, 1972; part III, p. 51; see footnote 8).

In other words, the image has lost all its picture-like qualities and has become a data structure meeting all the requirements on the form of a representation set forth in earlier sections. In fact, it can be put directly into one-to-one correspondence with a finite list of propositions. Thus the representation corresponding to the "image" is more like a *description* than a picture: There is nothing in the representation corresponding to the notion of "appearance." Similarly, "seeing the image" has been replaced by a set of common elementary and completely mechanical operations, such as that of testing for the identity of two symbols. The only reason that this could be done, of course, is that the I-space representation is in a canonical form in which tokens of a common type have a unique representation. Furthermore, only functionally relevant information is contained in the representation (e.g., size, coordinates of vertices, etc., are not represented, nor are spatial relations of diagonal elements). Recall that these were among the considerations which eliminated the concept of an image as an entity to be perceived.

Notice also that such a reformulation of the construct of imagery eliminates all reference to *perceptual* processes. But virtually all the informal definitions of imagery quoted earlier mention *perception* as being involved in imagery. Consequently, it is very tempting to conclude that Baylor's I-space representation has little to do with what the authors cited mean when they use the term "image." However, one could still argue that there is some functional similarity between the way in which the I-space representation plus its associated processes function in Baylor's theory and the way images function in the more informal theories (or the way the term *image* is used by subjects in the protocols). But notice that in interpreting processes in the system in terms of phenomenal descriptions, one is, in fact, redefining what the informal terms shall mean whenever they are used in an explanatory-theoretical capacity. And there is surely nothing wrong with this. But at the same time, it should be pointed out that words like image are undergoing a strange but essential transformation. They are being wrenched from their metaphorical context and are being given a role in a new formal system. This system then can shelter them from the excess meaning which they invariably carry in the informal theoretical context.

In the new formal context it also becomes possible — indeed compelling — to ask certain fundamental question which had been blocked when phenomena were described with words and phrases such as "image," "unitization," "spatial representation," "comparison of images," "reading an image," etc., serving as primitive (i.e., irreducible) constructions. It becomes possible to explore such formerly inaccessible questions as, What goes on in the "mind's eye?" *Why* can certain kinds of stimuli be more readily recalled? *Why* do certain mnemonics work? *Why* do certain classes of recall and performance tasks interfere or result in systematic confusions? The *why* in each of these questions could not be approached until the image metaphor is replaced by a fine detail information-processing model whose relation to the experience of imagery, by the way, is really quite a secondary matter. It is, in fact, significant that in this more formal model the experience of imaging has no causal role. It remains, at most, a source of ideas suggesting what processes might be required in the model.

University of Western Ontario

NOTES

* Several of the arguments appearing in this paper were first developed in the course of discussions held with Marvin Minsky and Michael Arbib. Also, the careful critical reading given an earlier version of this paper by Allan Paivio has hopefully led to a more careful and balanced presentation. I am grateful to these people for their help but am not so optimistic as to expect that they will agree with all the arguments appearing in the final draft.

[1] This claim does not depend on any assumptions about how information is encoded, so long as we hold that what is stored in some encoding of the particular stimulus token. Thus it holds for all types of encoding of the sensory pattern token including analogical ones such as holograms. It does not, however, apply to a view which has humans storing *procedures* which construct a representation anew from a finite set of primitive symbols each time a stimulus is encountered. Thus we are able to *discriminate* an unlimited number of stimulus patterns (e.g., numbers) even though we cannot store an unlimited number of such (encoded) patterns. This issue is discussed in Pylyshyn (1973) and in Fodor, Bever, and Garrett (in press).

[2] An even simpler way to deal with contradictions might be to adopt a proof procedure which blocked all derivations relying on contradictory premises. Although such an approach would prevent a knowledge base containing contradictory premises from degenerating, it would allow the base to contain both *p* and *not p*, which seems psychologically unreasonable. Instead it would seem more reasonable not to allow such a minimal pair but to allow certain cases of *p* and *q* even though *not p* is derivable from *q* (e.g., those cases in which the derivation exceeded the critical length).

[3] The distinction between a uniform proof procedure and a data-dependent one is subtler than this discussion might suggest. By adding selected theorems to the data base

or by ordering or otherwise marking the premises in some fashion, a uniform proof method can be made to behave very much like a data-dependent one although it might in practice be rather difficult to mold such a system to behave in some particular desired manner.

[4] In fact, the Gelernter system mentioned above, which is famous for its use of diagrams, never actually draws a diagram at all but merely constructs an internal representation of one. Furthermore, the representation need not be pictorial at all since, as Gelernter (1963) puts it "... the only information transmitted to the heuristic computer ... is of the form: 'Segment AB appears to be equal to segment CD in the diagram,' or 'Triangle ABC does not contain a right angle in the diagram (p. 139)." Such properties are not only propositional, but highly abstract (i.e., are true of a large set of possible diagrams). Admittedly, however, the "diagrams" do bear something of a type-token relation to the original description insofar as they achieve consistency and some degree of completeness by making arbitrary commitments with respect to certain aspects which are unspecified in the original description (e.g., approximate magnitudes).

[5] Quotations in the present paper are from an unpublished report with the same title which was issued in three parts by the Université de Montreal, Institute de Psychologie, 1971.

B. R. BUGELSKI

THE ASSOCIATION OF IMAGES

The current revival of interest in Imagery marked by the appearance of numerous books and articles suggests something more than a fad. The development of Cognitive Psychology in the last decade has made room for, if it did not require, images as components of cognitive processes. Those writers willing to incorporate images into their theorizing, however, have hastened to 'apply' the construct rather than to analyze it and, as presently viewed, images are somehow static 'parallel' processes relating to spatial organization of stimuli whereas (see Paivio, 1971) verbal processes are 'sequential'. Images are considered as responses (conditioned, or otherwise) to stimuli without a dynamic role in sequential thinking. Even when images are described as mediators in paired associate learning they are thought of as bridges between isolated pairs of words rather than as possible bricks forming a continuous road of imagery, or to change the metaphor, as bubbles in a stream of thought. Thus, in the use of mnemonic imagery one might use images to aid recall of a number of words but each response is still an isolated paired-associate. In my use of the "one-bun" technique (Bugelski, 1968) the fact that 10 items could be recalled in their ordinal position did not mean that *serial* or sequential learning had occurred; rather each item was paired with its own numeral stimulus as a paired-associate unit. Although an image of a bun might have been associated with an image of a battleship, the association presumably stopped there. For the experimental subject to say 'one-bun' subvocally, and 'battleship' aloud was taken to represent a complete sequence. The image of the battleship was presumed to do no more than to permit the labeling of the image with a word. The fact that an image of a battleship might generate new images in turn, e.g., of sailors, big guns, entire fleets, oceans, etc. was ignored. Such treatment of images as isolated, though mediating reactions, is curious in the light of the popular cybernetic view of words as functioning as both stimuli and responses, and subsequently, stimulus generators (see Saltz, 1971).

It is the purpose of this report to describe a more dynamic role of imagery as a sequential activity with one image leading to another as might

John M. Nicholas (ed.), Images, Perception, and Knowledge, 37–46.
All Rights Reserved.
Copyright © 1977 by D. Reidel Publishing Company, Dordrecht-Holland.

be supposed to occur in dreams of day or night time varieties or in any kind of train of thought. Unfortunately, to report on a sequence of imagery involves the use of words, and to generate any data we are forced to use lists of words as stimuli for some of the imagery generated by the experimental subjects. I shall describe some informal studies of imagery-chaining where, of necessity, I used serial learning of word lists as the data-generating task.

It is not expected of subjects who are exposed to a single presentation of a list of 20 or 40 randomly assorted words that they can report in sequence any great number of such words in a recall test. In the last decade we have become accustomed to the organizational operations of 'chunking', 'clustering' (Bousfield and Cohen, 1955), and 'recency' and 'primacy' effects (Glanzer and Cunitz, 1966); but, if subjects are asked to learn completely random lists of familiar words they are unable to do much more than to rely on some kind of stray associations that might be aroused. Probably no psychologist would expect subjects to report in correct serial order 20 words heard once unless some numerical mnemonic had been employed. Yet such a task is not beyond the capacities of literate human adults without any formal mnemonic crutch or 'system'.

What I am asserting is that it is a simple matter to report in serial order a large number of words heard once, granted an appropriate presentation rate and instructions to attempt to form an image to each word in relation to an image generated by the preceding word. Assuming that a given word arouses an image, this image – first a response – can now function as a stimulus for association with an image generated by the next word in a sequence, and so on, presumably indefinitely; certainly, the limits of such a capacity have not been tested. Thus I learned the following sequence in one-trial (read to me at a 5s presentation rate)[1]: car, book, spoon, dog, rose, pencil, moon, baby, soldier, house, camera, dime, flag, carrot, hat, river, knife, fire, wheel, cigar. The pattern of imagery was mildly amusing and, when taken as a whole (that is, embellished with connectives, verbs, etc.) resulted in a somewhat bizarre or fantastic story just as a dream might have a sequence of strange associations when considered as a whole. In the present instance, the word "car" prompted an image of an old-fashioned automobile, the next word, "book" aroused, in the context of the car image, an image of a textbook on auto repairs (the car suddenly became nonoperative), 'spoon' provoked an image of a need to mark the place in the book, and my spoon became a book-mark, the word 'dog' in the context of spoon called up an image of feeding the dog, emptying the contents of a can with the spoon. The dog, upon the mention of "rose" promptly wet the rose and the 'pencil' aroused the image of filling out an order blank for a new rose, and so on. Because each reporter will have his own private sequence of images there will be no correspondence between reports of different subjects. The various "stories" generated by different subjects would probably range over a number of dimensions such as simple-complex, logical-illogical, dull-interesting, etc. It might be speculated that any story-teller operates essentially in the same way, that is, reports in words, a stream of imagery. The only difference would be that the particular random words used here would not be controlling the story teller's operation. Another difference is that there are no time limits on story tellers and they are free to reject images that are not satisfying some ultimate goal or present purpose. The story-writer can wait until a 'good' image comes along.

It is apparent from the above report that there are no real controls over what images will occur — there is no prior selection or choice. We think what we must, although if the thinking is not solving problems or meeting needs, we think on. This is not to say that the thinker is completely helpless. He can reject images as unsuitable and wait for new ones, but this censoring operation is about all that is available for personal thought control. How this censoring is accomplished is as yet, unknown.

It is also apparent that the thinking consists of a sequence of images. A sequence of words never solved a problem — the word sequence *describes* a problem solution, assuming there is one; the actual solution may be the *sequence* of images which terminates with the matching of a need-fulfillment with its representation in imagery.

Because I had been my own experimenter up to this point I decided to see how well others could use imagery in serial recall and, accordingly, I used various groups of subjects in controlled situations as far as imagery instructions are concerned with a strong confirmation of my own findings. Even some 7th-grade children can recall 20 words in serial sequence after hearing them once. Some children do not recall very many but the average results show highly significant advantages of imagery subjects over verbal learning controls. The data have been published elsewhere, (Bugelski, 1974). and I will simply state the conclusions here, namely that people can recall a serial sequence of some rather astonishing length if they merely image each item mentioned in an integrated relationship with its successor.

Just to test the limits of this kind of function I asked some college girls to try to extend their capacities for serial recall. Starting with 5 subjects I had no problem with lists of 30 and 40 words. Then my problems started — the girls felt that the exercise was of no value to them, it had no interest — they were satisfied that they could go on more or less indefinitely but saw no point to it — neither did I, really, but I kept at it with one subject who reluctantly continued and could recite 80 words in succession without an error after one hearing. When I tried 100 words there were some hesitations on occasion but when I filled the gaps the recital continued. What I had succeeded in doing was to create a mnemonist out of someone who never thought of herself as possessed of a good memory, and, presumably, the same could be done with anyone. The only instructional feature of any merit in such training is to learn to have interacting imagery involving the items. What seems to work best is to image the first item having an effect on the second and vice versa — if the words "hat" and "table" are to be imaged, thinking of a hat *on* the table is of less value than having the hat under a table leg, getting crushed, while the table is being tilted. While the resulting image can be labeled as bizarre, it is not the bizarrity but the interaction that is the operating principle.

In ordinary thinking the succession of imagery is probably not so constrained — one image leads to another but the sequence might not be recallable backwards as it is in my 20 word studies. Because we are not ordinarily expecting a recall of our thinking we may wind up on occasion with thoughts we cannot trace, especially if no effort to image was attempted in some deliberate way. I am not going to digress into what might be labeled "unconscious imagery" but that might not be a bad label for Hebb's cell assemblies which is all I mean by the generation of an image in the first place. When there are no instructions to image at specific points and no need to label our thoughts we can presumably enjoy an entire stream of thought, emerge with a conclusion, and be unable to describe anything at all of what went on.

The use of imagery instructions at the present time is the only feasible way to get at the processes underlying thinking — such processes are presumably neural activities — they are not words — they are the partial revivals of the perceptual events that took place at some prior time and as such could include imagery of the appearance of words, books, pages, places on a page, and even the auditory aspects of spoken words. In this role as revived perceptual experiences words could indeed function in thinking and have a part to play — such a part would vary in its status from spear-carrier, to understudy, to lead, depending upon the importance of words in the original experience. Thus, I would conclude that words must be viewed as having a role in thinking sometimes but they cannot be regarded as the sole medium of thought which in my view would be the running-off of neural sequences representing a far greater background of experiences than verbal activities.

In writing this paper I found myself stopping on occasion and obliterating what I had just written because that is not what I wanted to say. Apparently I knew in some other way what I did want to say because I eventually settled for something. Certainly I knew what I did not want to say.

The trouble with talking about thinking is that we need to use words in talking and if words are also substantially involved in thinking then we find ourselves talking about what we are talking about. I think, i.e., perhaps, I say, we have made a first point, we must proceed slowly and with great care in this exercise.

The relation of speech and thinking has often been asserted, sometimes with great emphasis and conclusiveness, as in Watson's celebrated dictum that thinking is only "implicit speech" (Watson, 1919). The fact is that we all do talk to ourselves, implicitly and overtly, and the ubiquity of this observation had led many commentators to identify speech with a thought process. Because thinking is also commonly assumed to be the salient operation in problem-solving and because most of us mutter to ourselves in problem situations, the identification of thinking and talking acquires rather general support or acceptance. In the classroom the instructor can be quite convincing when he asks for a "mental" multiplication of some two-place numbers and the students readily recognize how they talk to themselves as they attempt the solution.

Should the word=thought formula not appear to be completely satisfactory we can always propose other implicit or explicit response mechanisms — eye movements, muscular tension, etc., to round out the picture but with the words, the inner speech, carrying the basic burden.

There is no point in denying the normal, common, and frequent exercise of speech in the daily goings-on of any normal human being. Speech, silent, or vocal is routine and general in our behavior. I am willing to accept such statements as "we are always talking to ourselves." The question, however, of what this talking means, is not answered by calling it thinking. It is still talking to ourselves and needs no other description or label, especially the label of "thinking" if it is not, in fact, equated with what might turn out to be some other process worthy of a name in its own right.

The moment we identify thinking with speech we deny thinking to a class of humans who cannot talk — to say nothing of problem solving animals. Pre-language children and deaf-mute children who have not yet acquired some sign language or vocalizations urged upon them, appear to learn readily enough in many situations that might be said to involve thinking. Certainly they behave appropriately in many situations and acquire so-called 'concepts' of tools, toys, furniture, edibility, etc. As they mature, according to Furth (1966), deaf children acquire the more abstract concepts of time, conservation, identity, etc.

In the last decade we have seen some psychologists attempting to train some of the great apes to use language either in the deaf-mute manner with sign language or through the use of symbols consisting of plastic forms. Such apes are reported to have formed a variety of sentences as well as a 'reading' capacity for sentences composed by their instructors. Did such apes not think until some kind of language had been acquired? When a chimpanzee forms a sentence indicating a specific desire to be assuaged by a specific individual can he be said not to have that desire prior to its expression in symbolic form?

There are some indications that normal human adults are, to some degree, at least, quite capable of thinking when their motor capacities are reduced and/or eliminated by curare — in their paralyzed state their speech mechanisms are also presumably inoperative yet they are able to solve problems, recall material, and know what is going on. Such research is necessarily infrequent and inconclusive. Yet more recent studies on split-brain patients suggest that the non-language side of the brain can function quite well in a variety of perceptual and memory tasks — we can hope for more data as such research becomes more developed. (see Gazzaniga, 1970).

So far I have suggested that the equation of thought to language is at least reduced in scope and power by our recognition of problem solving animals, mutes, pre-language children, language using apes, and clinical cases of paralysis and split-brains.

I wish to turn now to an area where language and thought are more closely identified in common parlance, but where the identity may be less than precise. I refer to the fact that many people are bilingual, even multi-lingual in various degrees of fluency. I am particularly interested in what is now a common case, that of the bilingual who learned one language as a child and subsequently, finding himself in another linguistic environment (by leaving home or through emigration) acquires a second language which becomes his usual communication medium. There are millions of cases of children of the foreign-born in countries like Canada, the United

States, Brazil, and Argentina. In New York City alone there are thousands of English speaking adults who learned Yiddish and some Hebrew in childhood but who no longer use these languages with either frequency or fluency. In my own case I learned Polish as a child but, in any serious sense, I have not used that language for about 50 years except on sporadic occasions. The question about such people is, and I'll make it personal, can I, or do I think in English most of the time? Do I think in Polish when speaking Polish or having interactions with Polish speaking people? Do I on such occasions think in English and translate into Polish? I can answer that last question in some part: I do not translate. The words, if they come, come directly.

But now we come to the point — I have frequently heard the claim that a bilingual thinks in one language and translates into another tongue for communication purposes. This presumably would apply to someone who is not competent in the communication language. Children learning a foreign language in school would presumably be in this category. The child learning Latin finds it easier to translate "from Latin into English" than "from English into Latin." I am not sure what this really means although I am confident about the statement. The fluent bilingual, presumably, should be able to think in either language. But here we have an immediate problem. If the fluent bilingual can think effectively in either language, presumably about the same problem or topic, does this not raise some question of what is the role of the language in the thought process? Does the sign-language teacher of the deaf think in signs? Does he think in words and translate into signs? Or does he have some underlying prior processes running off which he then *expresses* to others or to himself by a language medium? Is this also the case for the bilingual? What goes on in bilingual Morse Code Operators? He first identifies the language being transmitted and then proceeds to receive or report messages, to be sure, but what goes on in addition to hearing clicks?

Let us complicate the issue with the case of the multi-lingual innkeeper at a European resort who greets travelers from half a dozen countries in their own languages. If he is successful he is doing more than using words that are to be found in tourists' phrase books. His use of the words is based on different backgrounds, experiences, and what might be called an appreciation of the travelers' culture. Someone else hearing and using the same words would be engaged in a different operation — there would be different kinds of thinking involving the same words. Even with precise literal dictionary translations from one language to another there may be no communication if we simply rely on words. Do we ever know what someone is thinking from the words he uses in telling us his needs, problems, or concerns? What if he is lying? Is he thinking one thing and saying another?

Let me return to my own case. If a person can use a word to refer to some concrete object, say a piece of bread, in either of two languages and there is no mistaking of the denotation, i.e., the words will be recognized by lexicographers and translators as precise equivalents, we have a case where the same *thing* can be referred to by two words. But is it the same thing? I have tried to approach this question in two ways.

In the first approach I had some 30 English words translated into Polish. These words were sent to a colleague in Warsaw. The same words were sent to a former student in Manilla where they were translated into Tagolog. In each case the words were back-translated into English by others and then back into Polish and Tagolog to be certain of their lexical identity. The words were then subjected to a Semantic Differential test taken by approximately 100 American, Polish, and Philippino students. Not surprisingly, there were numerous significant differences in the ratings of the words on various scales. Differences occurred among all three groups of raters — Americans differed from Poles and Philippinos and Poles from Philippinos on different words. In many instances there were no critical differences but one could clearly conclude that, overall, the same common words (pig, house, street, door, etc.) did not mean the same thing to different language users. A word like "church", for example, received high positive evaluative ratings in Manilla, low ratings in Warsaw, and indifferent ratings in Buffalo, N.Y.

But these were different languages, used by different people, and one could argue that they still think in words even though the thought content might vary. This leaves open the question of what is the thought content.

My second approach was with the same person using the same words in different languages. This effort, thus far, is only in its beginnings and quite informal. I started with myself and jotted down 20 Polish words (names of concrete objects like bread, bed, school, street, etc.). The words were written down at random on different lines of a ruled page. The page was put aside for several days after which I looked at each word for about 5 sec. and reacted to it (whatever that may mean) with the "set" of associating each word with its successor using imagery as a mediator. When I reached the end of the list I tried to recall the 20 words in serial order. The task was accomplished with ease — no forgetting. As I mentioned earlier I have previously done this with lists of English words and have had no problem.

The successful recall of 20 serially ordered words in Polish was of no interest to me. What was of interest was the fact that the recall comprised a rather poor story of a childhood adventure, fictional to be sure, but comprised of possible experiences, set at about the age of 10. A few days later I went over the list in English and found myself developing a different story with the English equivalents although there were now, for me, intrusions from the Polish story. Perhaps if I wait a year or so I might be able to break that set and generate a completely different account. What I am sure of is that had I started with an English-worded list, the imagery involved would have been completely different.

Encouraged by this discovery, if it is one, I have informally approached some friends who happen to be, like myself, bilingual. So far I have merely checked my hypothesis with two or three words that I happened to know or acquired in Polish, Italian, Chinese, Yiddish, German, and in one case, Finnish — the Finnish words were supplied by the subject who was tested later. I informally asked my friends to report any imagery that occurs to them when I say the appropriate "foreign" language word. In every instance they have responded in terms of a childhood reference or experience. When I later check for English word reactions I get reports relating to current experience. My Italian respondent reacted to "milk" with a report of imagery about a carton of milk in his current home refrigerator. To "latte" he responded with a report of a *bottle* of milk in his childhood home. A Chinese graduate student has been probing the imagery of his Chinese fellow students and reports similarly about imagery related to China with a number of Chinese words and American experience with English words related to American experience like auto, apartment, bank, etc.

I am still in the process of generating a formal inquiry into this problem. So far all I know is that I do have a problem and a hypothesis. I am assuming that what happens when one engages in what would normally pass for thinking, there is a chain of imagery running off, each prompting a successor, with frequent recurrences of the same imagery now generating some other imagery, sometimes controlled by external situations and current needs, depending upon circumstances. When one indulges in reverie or reminiscence the degree of external control is minimal; when one is faced with a concrete situation, e.g., a dripping water tap, the situation determines the flow of imagery; if some other speaker is around, his words (and one's own) will also exert control over the imagery. But nowhere in the process do words function *per se* as the thought process. Even in situations where one thinks about words it is their sound or visual imagery aspects that comprise the thinking. Recently, Weber and Harnish (1974) reported a study in which subjects were required to describe the height of letters in words, i.e., they would be asked to say if the third or first letter, say, in the word "book" was tall or short. Subjects who were *shown* the letters on a screen did not respond as well as subjects who merely *thought* about them, i.e., were instructed "to image" the word.

From my observations on how students in other countries rate words that our English-speaking students rate otherwise as well as from my observations on how bilinguals function with their "old" language, I am convinced that the thought content is different. The difference appears to consist of the imagery generated by a particular word. One does not then think in German or Russian – one has German or Russian thoughts, i.e., imagery related to the times and places where those languages were functional – when they were used to name the perceptual experiences one was undergoing. I suspect that this would not be the case in any fluent bilingual who used his two languages daily from childhood – for him the imagery would be the same, no matter which language he happened to be using, and, in all probability the images would be relatively more current, i.e., from recent experience rather than from childhood.

I will conclude with the remark that my use of the term Imagery does not imply anything beyond the activation of some neural activity either through external stimulation or through the internal activation of one neural response by a prior one. My orientation about imagery goes back to my first reading of Professor Hebb's *The Organization of Behavior* where, I am quite certain, there was no reference to the operation of words, as such, in the nervous system. At the same time I am not trying to suggest that words play no role or even a small role in the control of thinking; that would be manifestly absurd. All I hope to do is to try to talk about thinking instead of about talking.

State University of New York at Buffalo

NOTE

[1] This presentation rate is sufficient for some subjects; an 8 s interval might provide more success with more subjects. See Bugelski, Kidd, and Segmen (1968).

ALLAN PAIVIO

IMAGES, PROPOSITIONS, AND KNOWLEDGE

The concept of imagery has been linked historically to both memory and thought. The memory side apparently originated in Western culture as a mnemonic technique that was invented by Simonides about 2500 years ago. The method was promoted by orators as a way of memorizing speeches, by educators who advocated the technique as a learning device during the middle ages, and by philosophers such as Bruno who wanted to unify all knowledge within memory through the organization of images. Imagery as a theory of thought also goes back to antiquity. Aristotle wrote that it was impossible to think without mental pictures, and used the images of mnemonics to illustrate his statements about imagination and thought. Later, the British empiricists equated images with ideas in their theory of association. Images were subsequently viewed as basic units of consciousness in the introspective psychologies of Wundt and Titchener. Finally, today, both functions of imagery are given a prominent place in some theories of memory and cognition, but with important changes in the assumptions concerning the nature of the concept. I will discuss those differences. presently.

1. RECURRENT CRITICISMS OF THE IMAGE

One of the most striking facts concerning the history of the concept is that it has been repeatedly criticized and opposed and yet the critics never quite succeeded in burying the image. I believe there are important contemporary lessons to be learned from the history of the controversy, so I will remind you of some of the critical highlights. In connection with the role of imagery as a memory aid in public speaking, Quintilian noted the difficulty of finding images to represent function words and abstract words. He also asked 'How can such an art [that is the imagery mnemonic art] grasp a whole series of connected words?' and further 'will not the flow of our speech inevitably be impeded by the double task imposed on our memory? For how can our words be expected to flow in connected speech, if we have to look back at separate forms for each individual word?' Quintilian apparently recognized the problem of representing abstract ideas in the form of images, and the overloading of memory that would result if speech were mediated by conscious inspection of detailed, reproductive images. These problems have been repeatedly expressed since then. Locke and Berkeley, you will recall, had difficulty with the imagistic representation of general ideas: Locke argued for some kind of abstract of composite imagery, whereas Berkeley believed that imagery is always individual and concrete. Again in 1958, Roger Brown rejected imagery as a theory of meaning precisely on the grounds that images could not represent abstract or general meaning. Finally, John Watson opposed imagery because it was so clearly tied to the concept of consciousness and the introspective approach that he abandoned.

Each of these criticisms has been echoed by contemporary proponents of computer simulation models of memory and cognition, especially by Zenon Pylyshyn in his critique of mental imagery in the *Psychological Bulletin* (1973) and John Anderson and Gordon Bower in their recent book on *Human Associative Memory* (1973). These critics argue that subjective imagery is unsatisfactory as a primitive explanatory construct because what is available to conscious inspection may not be what plays the important causal role in remembering or other psychological processes. The term also implies that what we record in memory are raw, unanalyzed, textured details which must be scanned perceptually at a later date in order to retrieve meaningful information. This, says Pylyshyn, places 'an incredible burden on the storage capacity of the brain' (p. 9) because there is no limit to the variety of possible sensory patterns that can be distinguished. In addition, these critics raise again the ancient objection that such images, being specific and detailed, cannot represent general or abstract knowledge.

The opponents in each era also proposed alternatives to imagery theory, based either on natural language or logical descriptions. The alternatives never proved entirely satisfactory in the earlier history of the controversy because they could not explain the phenomenal experience of imagery nor its functions in cognition. This was true of the Watsonian suppression of imagery, which has been countered in the last decade by the weight of empirical evidence showing the insufficiency of verbal explanations of performance in traditional experimental tasks for studying human memory. I will argue later that the same fate awaits the propositional or computer simulation models of long term memory and cognition as they are presently formulated, but first I will try to show that the traditional criticisms of imagery are currently unjustified.

2. THE METAPHORICAL BASIS OF THE CRITICISMS

The criticisms of imagery have always been directed at a particular metaphorical model of the mental image. Its history clearly illustrates how our thinking can be misdirected by the implications of a metaphor when taken too seriously. The metaphor has varied in its specific form but it has always been expressed in terms of the available technology for recording visual information. Thus inner images have been compared at different times with engravings on wax tablets, painted portraits, photographs, and motion pictures. Such representations are reproductions of reality. The information they contain is limited to the recorded details and the processing of that information requires a conscious viewer to scan the image. It follows by analogy that mental pictures, too, must be reproductions to be viewed by the mind's eye. The knowledge contained therein would be what has been passively recorded and its retrieval would depend on the clarity and detail of the impression or picture. The metaphorical model has left its mark on our everyday language. We speak of our memories as *vivid, faded,* or *erased,* depending on whether the original perceptual experience was *impressive* or *unimpressive.* Impression, vividness, fading — such words make sense only in terms of some form of the wax tablet model of memory. Moreover, the metaphor influenced the earliest empirical research on imagery. Galton's famous Breakfast Table Questionnaire and subsequent versions of it were explicitly intended to measure individual differences in subjective vividness of imagery.

3. INAPPLICABILITY OF THE CRITICISMS TO CONTEMPORARY THEORIES OF IMAGERY

The wax tablet or picture metaphor is clearly open to the kinds of criticisms that were directed at it over the ages. Today, however, they are largely directed at a straw man because no major imagery researcher accepts the metaphorical view as a working theory. It is very important to be clear about this point because otherwise our future is written in our past and we are doomed to keep repeating the logical errors and the endless debates. I will therefore try to be specific in regard to each of the major misconceptions.

3.1 Role of Consciousness and Introspection in the Definition of the Image

Consider, first, the role of conscious experience, which is associated with the wax tablet model because such a representation requires a conscious viewer. This is no longer used as a basis for defining the functional attributes of imagery and for this shift we must be grateful to John Watson. If used at all, introspective evidence primarily supplements other information in supporting the inference that imagery is *involved* in task performance. Subjective reports might also be used as an informal source of research ideas but they are not used to *define* the functional properties of imagery. In fact, introspection clearly failed to reveal the functions of imagery in the earliest research on the concept. First, the Wurzburg psychologists demonstrated that thinking could occur without conscious content in the form of either imagery or inner speech. Second, the early research on imagery showed that individual differences in the vividness of experienced imagery was a poor predictor of memory performance and some recent attempts to use this approach have also been relatively unsuccessful. All of the more useful evidence on the nature of imagery has come from objective methods in which the involvement of imagery in a given task is encouraged by the use of relevant stimulus materials such as pictures or concrete words, by giving subjects instructions to image, and so on. These procedures alone do not reveal the functions of imagery nor are they intended to do so. The inferences about function are based instead on the measurable relationship between the demands of the task and how subjects perform on it. The results of more than a decade of memory research by many people who have used such methods, including Bugelski, Bower, and my students and former students, permit us to infer that imagery aids learning in a wide variety of tasks, that it does so partly by integrating or unitizing multiple units of information in memory, and so on. The developmental studies of imagery by Piaget and others are similarly based on children's performance on objective tasks that may require only static reproductive memory of the situation, or go farther and require mental manipulation of information in order to solve the problem. Again, Roger Shepard's recent work on the rotation of mental images has involved completely objective reaction time methods which not only confirm the subjective impression that we can rotate mental images, but also provide a means of measuring the speed of such rotation. If in addition subjects say that they used imagery in such tasks, it supports the inference that a particular mental process occurred but it is the objective performance that reveals precisely what the process can do.

IMAGES, PROPOSITIONS, AND KNOWLEDGE

3.2. Images are not Passive Reproductions

The second point is that contemporary researchers do not assume that images are simply passive and highly detailed reproductions, like photographs. Imagery is viewed instead as a dynamic process more like *active perception* than a passive recorder of experience. Thus Piaget and Inhelder find the origins of imagery in imitative behavior and perceptual activity, and the resulting process includes a motor component which enables it to perform transformations on remembered information, and to anticipate the outcomes of concrete operations that have never been performed by the subject. Hebb and Skinner similarly emphasize motor processes in imagery — Skinner of course discussing the phenomenon in terms of conditioned or operant seeing. Bugelski, in an article on the definition of the image, has also suggested that imagery is an active process and not a thing in the sense of an object to be viewed. My own theoretical ideas include the conception of imagery as a dynamic symbolic system capable of organizing and transforming the perceptual information we receive, and much of our research has been directed at testing such views. Today's theorists sometimes talk in terms of the picture metaphor as everyone else does, and as I will continue to do here, but this does not mean that the literal implications of the metaphor are introduced into the predictive theories.

3.3 Image Memory Does not Require 'Unlimited Storage Capacity.'

The notion that image memory implies unlimited storage capacity is a corollary of the view that images are detailed productions, so it, too, is a misconception. No one today denies that perception and memory are selective and incomplete, nor that reconstructive or generative processes of some kind are involved when perceptual information is remembered in an imagistic form. There is no reason whatever to assume that such imagery must record experience in all its raw detail. Precisely how much knowledge can be so represented is entirely an empirical matter.

3.4 Imagery and Abstract Knowledge

Finally, we have the often-repeated criticism that images, being specific, cannot represent general or abstract knowledge. There are three answers to this. First, within limits, imagery might involve considerable abstraction in the sense that the representation includes only the essential attributes of a class of things. This is abstraction in much the same sense as a line drawing or a caricature is abstract or schematic. This has been proposed by many psychologists and I have not yet seen any good reason to reject it as a partial answer. Second, many psychologists have suggested that general ideas can be concretized in the form of specific images. For an individual, a specific exemplar and its image somehow functions as the prototype for a general class. The arguments against the prototype view generally involve the misconception that there must be a one-to-one relation between the concept and a particular referent image. This need not be so. Images can be regarded as associative reactions, the occurrence of which depends, not only on the particular stimulus to which one reacts, but also upon one's past experiences with particular referent objects and their labels and upon the situational context in which the stimulus occurs. Thus the word *dog* can arouse different images in different people and in a given person on different occasions, and the image is somehow representative of the specific meaning of the word for that person at that time. Moreover, the alternative referent images may constitute a hierarchy that can vary in number and relative availability of the alternatives, so that a particular kind of imagistic reaction is most probable for a given concept. Others have independently presented similar views and have even provided some clear empirical support for them. Eleanor Heider Rosch (1974) recently reported evidence from a series of studies to the effect that people represent conceptual categories mentally as analog prototypes that are the best examples of the categories. Richard Anderson and Barry McGraw (1973) concluded similarly on the basis of recall data that people use images of highly probable exemplars to represent the meanings of general terms. Thus the effective representation of ANIMAL is more likely to look like a *dog* than a *squirrel*.

The third comment on the issue is that imagery need not carry the entire burden of accounting for abstract concepts. It may be that highly abstract concepts are processed entirely or primarily verbally, with little involvement of the imagery system, or that imagery and verbal processes interact in some complex fashion which remains to be worked out by appropriate research. The general point at the moment is that, far from being ruled out, imagery is becoming increasingly plausible as a representational basis for general knowledge — increasingly plausible on empirical grounds.

4. PROPOSITIONAL REPRESENTATIONS

I turn now to the abstract linguistic or propositional alternatives as proposed by the critics of imagery. First I will give some examples of such approaches and then I will describe some of their inadequacies as I see them.

This general approach, like imagery, has an old history. The following system, introduced in France by Peter Ramus in the 16th century, was part of an attempt to simplify teaching methods largely through the use of better ways of memorizing subjects. Ramus explicitly rejected the old imagistic art of memory and substituted a method called "dialectical order." A subject was set out in schematic form moving from general to more specific aspects in a series of dichotomized classifications. An example is shown in Figure 1. Note that the schematic presentation consists essentially of a hierarchical tree structure with binary branchings at each node. From this schema, the ramifications (literally, the branchings) of the subject were to be memorized. Memorizing from such a layout involves spatial visualization, but the content is clearly abstract and verbal in form, consisting of class labels under class labels, ranging from the generic to the specific. It is the 16th century epitome of an abstract, logical tree of knowledge.

Ramus also had a vision of the atomic units underlying such structures. According to a Ramist biographer "Ramus had developed the habit of regarding everything, mental and physical, as composed of little corpuscular units... [He] tend[ed] to view all intellectual operations as a spatial grouping of a number of these corpuscles into a kind of cluster, or as a breaking down of clusters into their corpuscular units. The clusters, once formed, can be regarded also as corpuscles which in themselves admit a further combination and which form still further clusters of clusters (Ong, 1958, p. 203)." I am sure this has a familiar ring to many of you!

Ramism has its contemporary analogs in a variety of network models of semantic or long term memory, which assume that the structural units of memory are either like words or unspecified semantic atoms which in turn are organized into such higher-order structures as logical propositions. The most influential of these have been developed as computer simulation models (e.g., Anderson and Bower, 1973; Collins and Quillian, 1972; Rumelhart, Lindsay, and Norman, 1972). These and other current models of long term memory differ in important details, but the relevant point here is that they are similar in their emphasis on a common representational format for linguistic and nonlinguistic information.

A look at Anderson and Bower's schematization of human associative memory (HAM) will help us appreciate the overall structure of such a model. Figure 2 shows those features of the model that are most relevant here. Note especially that linguistic and nonlinguistic perceptual information come in separately through a sensory stage into separate parsers, but

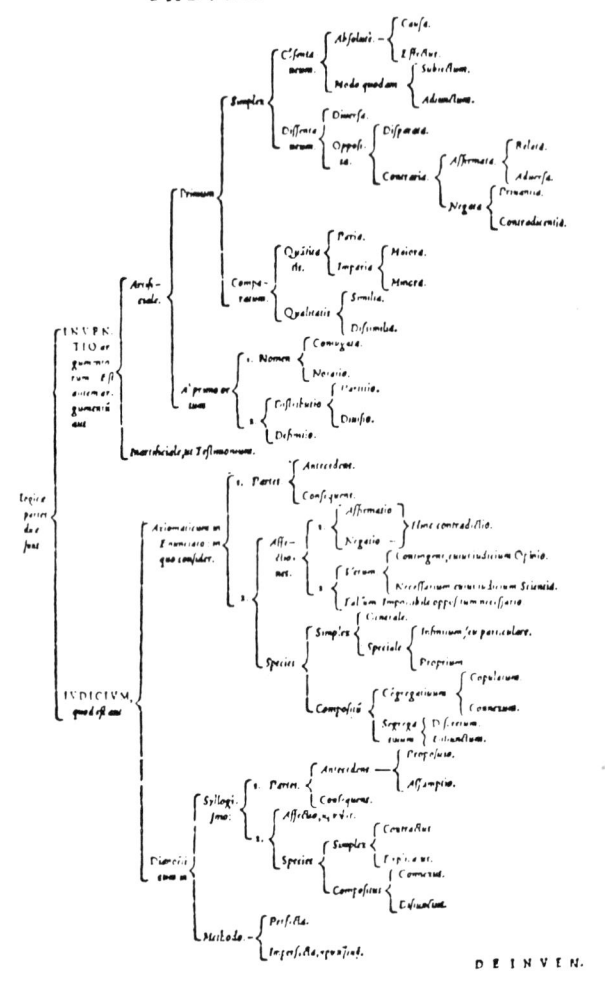

Fig. 1. An example of the branching tree structure resulting from the method of dialectical order as advocated by Ramus.

IMAGES, PROPOSITIONS, AND KNOWLEDGE

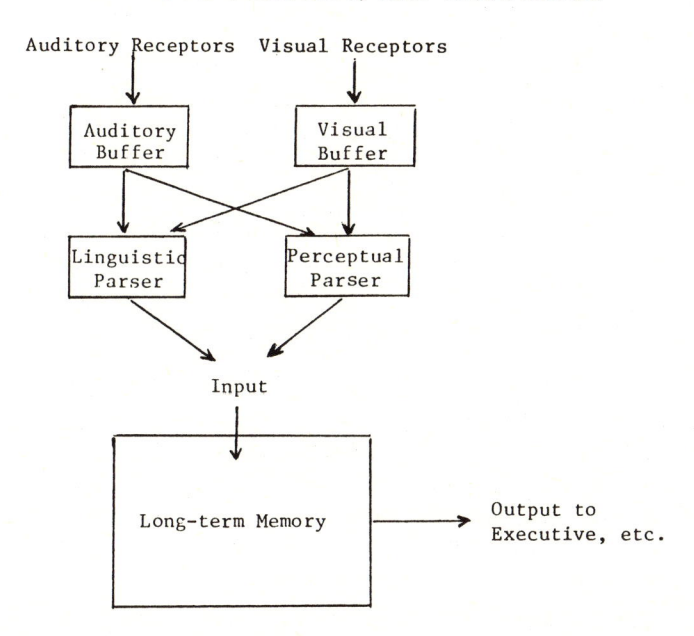

```
Auditory Receptors    Visual Receptors
       ↓                     ↓
  ┌──────────┐          ┌──────────┐
  │ Auditory │          │  Visual  │
  │  Buffer  │          │  Buffer  │
  └──────────┘          └──────────┘
       ↓   ↘          ↙    ↓
  ┌──────────┐          ┌──────────┐
  │Linguistic│          │Perceptual│
  │  Parser  │          │  Parser  │
  └──────────┘          └──────────┘
           ↘          ↙
            Input
              ↓
  ┌─────────────────────┐
  │                     │         Output to
  │  Long-term Memory   │ ──────▶ Executive, etc.
  │                     │
  └─────────────────────┘
```

Fig. 2. Schematization of HAM showing the input of linguistic and nonlinguistic (perceptual) information to a common long-term memory component. Adapted from Anderson and Bower (1973, Fig. 6.1 and accompanying text, pp. 137–138).

then the output of these parsers go into a common memory system. This memory box contains all the long term memory information accumulated from experience. The information can be retrieved and operated on by various executive processes that we need not consider here. The main point is that there is a single long-term memory system for both linguistic and perceptual information, and that both are represented in the same kind of propositional format. The model has been applied primarily to the processing of sentences, but it is assumed that perceptual information would be handled in the same way (see, for example, the proposed progression from percept to description in Figure 8.8 of Anderson and Bower, 1973, p. 214).

I will now discuss three general problems with this kind of approach to cognition.

4.1 Descriptions are Inadequate as Primitive Explanatory Concepts

First, the assertion that knowledge is represented in the form of descriptive propositions is not an explanation, it simply begs the question of the psychological nature of such knowledge. The nub of the problem is that the concepts that enter into propositional statements must themselves be defined. This has usually been done in one of two ways in the descriptive models. One is the dictionary approach in which every concept is defined verbally in terms of other concepts in a closed network of associations. This is essentially the procedure that has been adopted by Collins and Quillian, as well as their successors (see, for example, Figure 9 in Rumelhart, Lindsay, and Norman, 1972 p. 227). In discussing the distinction between words and ideas in their model, Anderson and Bower similarly assert that, "the idea nodes are essentially nameless entities that acquire their meaning from the configuration of associations into which they enter with other ideas (p. 207)." Such a definition is completely circular and unsatisfying because it never leads to any psychological knowledge other than words or "nameless entities" rather like Ramian corpuscles.

The other approach is to introduce terminal elements to break the infinite regress. Katz and Fodor (1963) in their proposal for a linguistic semantic theory, attempted to do so by postulating *distinguishers* to represent whatever is semantically unique to a word. They recognize that the terminal atomic feature, the distinguisher, was not a psychological unit representing knowledge of the world, which they deliberately excluded from their approach. The ultimate nature of the distinguisher and other semantic features were left as a problem for psychology. In current psychological models of semantic memory, the problem is handled by postulating perceptual features or simple ideas as the terminal atomic units. Anderson and Bower, for example, assert that "we share the sensationalist bias of past associative theories in believing that the mind has been shaped through evolution to encode perceptual information, and that all inputs to memory are basically perceptual descriptions (p. 154)." However, computer programs are presently very unsatisfactory for handling perceptual problems like scene analysis and these necessarily remain pure assumptions in the model, as they frankly admit when they say that their theory is "sensationalist in spirit rather than detail, because these sensationalistic beliefs have not been realized by specific theoretical proposals (p. 187)." Thus it is completely gratuitous to assume, as they do, that linguistic and perceptual memories alike are propositional descriptions constructed from the same kinds of perceptual atoms. Assigning labels to theoretical entities does not reveal the psychological reality that they are supposed to represent.

In this connection I am reminded of a marvelous allegory in Ibsen's Peer Gynt. Ibsen's hero wanted to learn what the essence or core of the onion was like, so he proceeded to peel off its layers one by one. He continued to do so until the last layer had been peeled away and nothing was left. There was no core, no essence, to the onion. It seems to me that this aptly represents the present state of the descriptive approach to the definition of concepts. It strips away the psychological reality, the onion, and leaves us with a verbal promise that the onion will be reassembled by a construction program yet to be invented.

4.2 Propositional Descriptions do not Solve the Problem of Memory Load

My next point concerns the problem of memory capacity. How are we to explain the enormous amount of knowledge we retain and can use? This is as much a problem for the propositional theories as it is for imagery. Some kind of generative approach has often been suggested as a parsimonious solution to the problem. Pylyshyn's version of it is that "human store *procedures* which construct a representation anew from a finite set of primitive symbols each time a stimulus is encountered. Thus we are able to *discriminate* an unlimited number of stimulus patterns (e.g., numbers) even though we cannot store an unlimited number of such (encoded) patterns (p. 9)." The example is misleading, however, because it apparently refers to perceptual discrimination of numerical stimulus patterns from each other rather than to the amount of discriminable information in long term memory. Perceptual discriminations can be done on the basis of a simple physical comparison between two numbers as stimulus patterns. No large storage capacity is required in this case, nor is it required in the case of the generation of novel numerical responses, as in the multiplication of two numbers that one has never multiplied before. This can be done on the basis of a relatively small number of rote procedures applied to a small number of numerical units in memory.

The number of potential memory discriminations that we can make on the basis of our everyday knowledge of the world poses a problem of quite a different magnitude. Consider, for example, the identification of all the faces that we know. We must have stored whatever information is necessary to recognize a particular face as belonging to a particular individual rather than to the thousands of others that we may know. The generative approach says that we in effect construct the face from elementary perceptual features through the operation of some kind of construction rules. But this does not reduce the strain on memory because the rules relevant to the memory identification of Joe Smith must be different from those involved in the recognition of Jim Jones and every other person that we can identify. This difference must be represented in memory in some way. The problem can be more concretely illustrated by the number discrimination problem. Suppose I perform a thousand multiplication problems so that I have a thousand different numerical products, and then I associate a name with each as though it were someone's telephone number. Then I repeat the procedure until I can supply the name when given the number, or generate the number from the name. To accomplish this feat, I would either have to remember the particular procedure I went through to construct each number and its associated name, or store both as some kind of template. Surely each combination of elementary features and construction rules would take at least as much memory space as a separate representation for each name and number. Given this problem, the generative model simply has no advantage over a wax tablet or template model as far as memory capacity is concerned. The same problem applies to propositional network models of long-term memory, as I will show later in connection with a specific research example.

4.3 Restrictive Implications of the Propositional Models

Finally, in my view, the most serious shortcoming of the computer-based propositional models is that they impose arbitrary limitations on the interpretation of imagery and other cognitive functions. These limitations stem from the assumption that all knowledge is represented in the same conceptual format in long term memory. This means, theoretically, that all perceptual, imagistic, and linguistic information is put into a one-to-one correspondence with the same propositional base. Thus Pylyshyn, in his critique, asserts that what are called mental images contain *only* the kind of information that can be expressed or potentially expressed in some formal, proposition-like language. Drawing an analogy between image and physical object as models, he writes, "so long as the physical object is being used *as a model*, all inferences drawn from it were entailed by the propositional representation ... from which it was constructed. Thus the model (that is, the image) introduced no new information although it serves the invaluable function of making what was implicit in the description more explicit, accessible, and manipulable (p. 19)." Note the implication of this approach: instead of leaving the nature of imagery open to inquiry, the issue is closed off by the prescriptive assertion that images are generated *only* from descriptive propositions and contain *only* the information in such descriptions. But this limitation is forced on us by the underlying computer metaphor and not necessarily by psychological facts. Pictures generated by today's digital computers are constructed bit by bit from informational units in formal descriptions so that a one-to-one correspondence is inevitable, much as in the relationship between geometry and algebra. This limitation is extended to imagery, as explicitly stated by Pylyshyn in a comment on Baylor's (1972) simulation model of imagery: in such an approach the image "can be put directly into one-to-one correspondence with a finite list of propositions (p. 22)." Thus a complete dependency is assumed between proposition and image or picture.

This approach would be theoretically innocuous if it simply means that there is an abstract substrate to imagery and it contains the information necessary to generate the image. This is obviously true in some sense — everyone would agree, for example, that there is a neural substrate to imagery. However, the assumption is redundant in a psychological theory, since all we can get at empirically are the behavioral and experiential expressions of the underlying system. Thus a theory of imagery is at the same time a theory of its substrate as a psychologically functional system.

On the other hand, it is not theoretically innocuous to assume that the system which interprets scenes or generates images also interprets or generates language, since we now have a one-to-one relation between linguistic description, proposition, and image. What is most restrictive here is the implied one-to-one relation between language and imagery. It means that we should be able to imagine only what we can express verbally, and vice versa. Moreover, we would never be quite sure whether our knowledge of the world came from actual experience with the world or from reading or hearing about it, since both kinds of information become neutralized in long-term memory. Such an impoverished view of mind does not square with commonsense nor with research evidence, as I hope to show presently.

An alternative possibility within the same framework would be to assume a common substrate for language and imagery, but to add that the common system contains different regions or differentiating tags corresponding to linguistic and imaginal information. In that case, the propositional model would become formally indistinguishable from a dual coding theory which simply assumes that imagery and language involve separate systems. I shall now outline such an approach and then turn to some relevant experimental data.

5. A DUAL CODING APPROACH

The proposed approach is an alternative to both the wax tablet model of imagery and the common coding theories of knowledge. The most general assumption is that verbal and nonverbal information are represented and processed in independent but interconnected symbolic systems. A second general assumption is that the nature of the symbolic information differs qualitatively in the two systems. Specifically, one system is specialized for representing and processing information concerning nonverbal objects and events in a rather direct, analog fashion. Since imagery is a unique expression of its functioning, it is convenient to refer to it as the imagery system. The other is specialized for dealing with linguistic units and generating speech, so let us call it the verbal system. The qualitative distinctions extend to the way information is organized in the two systems. The imagery system organizes elementary images into higher-order structures so that the informational output of the system has a synchronous or spatial character, whereas the verbal system organizes linguistic units into higher-order sequential structures. Finally, rather than being static representations, both systems are assumed to be capable of reorganizing, manipulating, or transforming cognitive information. When such changes occur, they are assumed to be governed by the structural features of the information in each system. Thus dynamic visual imagery involves transformations of such visual and spatial attributes as color, size, location, and orientation of imagined objects, whereas verbal transformations involve rearrangements of the sequential ordering of words and other linguistic units.

This approach has psychological implications relevant to the critical issues and alternative approaches discussed earlier. First, there is no implication in any of this that the effective process must be conscious. The two processes are defined in terms of their functional properties and experienced imagery and language are simply conscious expressions of the activity of the underlying systems. Consciousness *per se* does not define *what* the systems can do any more than it does in the case of the propositional models. Second, the dual coding model implies *independence* of the two systems, so that the content of information in one system is not necessarily dependent on or predictable from the other. Thus my knowledge of the world includes information that has never been expressed verbally to me or by me, such as the appearance of my home or my office, and a good deal that is impossible or difficult to describe. Conversely, much of my verbal knowledge has no direct nonverbal counterpart. Independence also implies that one system can be active without the other, or both can be active concurrently but not necessarily processing the same conceptual information. For example, I can observe what is going on in this room while talking on a topic quite unrelated to the contents of the room. I even have the impression that I can talk while imagining my living room or cottage in some vague way. These implications do not arise from the propositional models, for they assume a dependency relationship rather than independence of verbal and nonverbal processes.

Next, dual coding assumes rich but partial interconnectedness between the two systems rather than a one-to-one correspondence. Interconnections are of course necessary to capture the idea that nonverbal information can be transformed into verbal, or vice versa: objects can be named, names can arouse images, or such exchanges can occur between images and words entirely at an implicit level. The idea that the interconnections are only partial means that there is limited access between systems. One system can elicit activity in the other only at points where interconnections have been established. Once a system has been activated, however, further processing can go on entirely within that system relatively unconstrained although potentially modifiable by the activity of the other.

The assumptions of independence and interconnectedness have implications in regard to the processing of pictures or scenes as compared to descriptions, and concrete as compared to abstract language. Pictures or scenes presumably activate the imagery system rather directly, whereas descriptions do so only indirectly; conversely, words or descriptions set up activity in the verbal system relatively directly and elicit images indirectly. In regard to linguistic concreteness, the connecting pathways from the verbal to imaginal representations are assumed to be relatively direct in the case of concrete nouns and more indirect in the case of abstract nouns and word classes other than nouns. Thus an abstract word like "religion" may arouse imagery by first activating some concrete verbal associate like "church." Similarly, general terms like "animal" might arouse imagery only by first being translated into a specific term such as "dog." However, I must emphasize again that none of this implies a one-to-one relation between verbal representation and a particular image even in the case of the most concrete and specific item. Instead, a word or phrase presumably arouses different images, or an object or event different verbal descriptions, depending on one's past experiences and the context in which the reactions occur.

Finally, the dual coding model assumes that the representational units of nonverbal information are wholistic perceptual analogs or images whereas verbal representations are functionally more like words or other components of inner speech, bearing no resemblance to things. Propositions, as described in the current computer simulation models, resemble structures generated from such linguistic units, but they do not resemble things or images.

6. SELECTED RESEARCH EVIDENCE

I turn now to some specific evidence on the theoretical alternatives. First I will simply mention some of the kinds of data that support the idea of two independent cognitive systems, then I will describe some recent studies that show the analog nature of images and bear on the issue of memory capacity.

6.1 Evidence for Dual Coding

Neuropsychological studies provide rather direct evidence that linguistic and nonlinguistic information are processed by different symbolic systems. The evidence comes from studies of patients with localized damage to one or other side of the brain, patients in whom the two hemispheres have been surgically separated, and from studies in which verbal or nonverbal stimuli are presented to the right or left hemisphere of normal subjects. The evidence generally suggests that the processing of linguistic memory information is in some fundamental sense localized in the left hemisphere for most people, whereas certain kinds of spatial and nonverbal skills are specific to the other hemisphere, usually the right. Such evidence suggests that the two kinds of information are stored and processed independently to a significant degree. The findings are not readily explained by the view that verbal and nonverbal information are represented in a common long term memory system.

Individual differences in human abilities also provide evidence for separate systems. Specifically, verbal and nonverbal abilities are factorially independent when measured by relevant tests. Thus, in Guilford's (1967) research, the figural abilities that are measured by the Minnesota Paper Form Board, Space Relations, and nonverbal memory tests comprise a set of factors that are quite independent of verbal abilities as defined by associational fluency and numerous other tests of symbolic and semantic skills. We have similarly found that spatial manipulation and questionnaire tests of imagery are independent of various verbal ability tests. Why should we find this general independence of verbal and nonverbal abilities if the two kinds of knowledge are represented in a common symbolic format? Again, the evidence seems more consistent with a dual coding view.

A third kind of support evidence for dual coding comes from studies comparing pictures and words or imagery and verbal coding instructions in memory tasks. We have found, for example, that the recall of pictures and their verbal labels is independent in the strong statistical sense that the two kinds of stimuli have additive effects on recall. Loosely speaking, repeating a word as a picture or vice versa doubles recall, whereas repeating a picture twice or a word twice in succession does not double recall. We have also found that implicit naming and mental images aroused by instructional cues are similarly additive in their effect. Other investigators have found that pictorial and name information can be independently forgotten.

Several other kinds of evidence support dual coding and not the common code theories, but the ones I have mentioned suffice to make the point.

6.2 Evidence for Qualitative Distinctions in the Two Codes

I turn now to evidence that bears particularly on the analog nature of images. Roger Shepard's work on the rotation of mental images is perhaps the best known. The mental phenomenon is illustrated by the following example. Several years ago I asked my youngest daughter to imagine a capital N. She said "Okay" and I said "Now tip it over on its side." I asked, "What do you see?" and she immediately replied "I see a Z!" Shepard and his collaborators have been able to measure the speed of such mental rotations using reaction time methods. If you haven't yet read any of the reports on these elegant experiments, I recommend that you begin with Lynn Cooper and Roger Shepard's (1973) chapter in a recent book on visual information processing edited by Chase. The significance of these experiments for the problem of knowledge is that they provide experimental evidence that we know how things appear directly, in a visual sense, as well as how they appear when they have been transformed in some way. The results tell us that the mental representations must contain a dynamic component that functions internally much as perceptual-motor processes do when applied to the concrete world of objects and events. The idea that this analog process is imagistic in nature is supported by the fact that subjects consistently report using imagery in these tasks. Even Anderson and Bower concede that this kind of competence may be beyond their proposition-based computer simulation model and therefore a positive argument for imagery.

The analog nature of knowledge is even clearer in the next example. When people are asked which of two stimuli is larger, the reaction time for the decision varies inversely with the size difference. The greater the difference in the lengths of two lines, for example, the faster the decision time. Robert Moyer (1973) reported a study recently in which subjects compared the sizes of named animals from memory. The animal names had been previously ranked according to the relative sizes of the animals themselves. The experimental subjects were visually presented the names of two animals, such as *frog-wolf*, and were required to throw a switch under the name of the larger animal. The interesting result was that the reaction time for the choice increased systematically as the difference in animal size became smaller. Since this function is similar to the one obtained when subjects make direct perceptual comparisons, Moyer argued that subjects compare animals names by making an "internal psychophysical judgment" after first converting the names to analog representations that preserve animal size. Smaller size differences between animals presumably are represented as smaller differences between the internal analogs, and a resulting decreased discriminability if reflected in increased reaction times.

We have recently extended Moyer's study in various ways in order to determine whether the analogs are like images or at least wholistic visual memory representations of some kind, rather than verbal entities or descriptive propositions. First, we extended the range of named objects so that the list included inanimate object pairs, animals paired with objects, as well as animal-animal pairs. The result was that we replicated Moyer's general function for all three kinds of comparisons, as can be seen in Figure 3.

The experiments included controls that pretty well ruled out any kind of verbal response bias as an explanation of the results. Could semantic feature or propositional network theories account for the data? Consider how Anderson and Bower's human associative memory model might handle such a problem. Each relative size difference would be represented in a propositional tree consisting of a subject and a predicate, together with a relation "larger than" (cf. HAM's representation of the relation "taller than," Anderson and Bower, 1973, Figure 7.3a, p. 158). The propositional

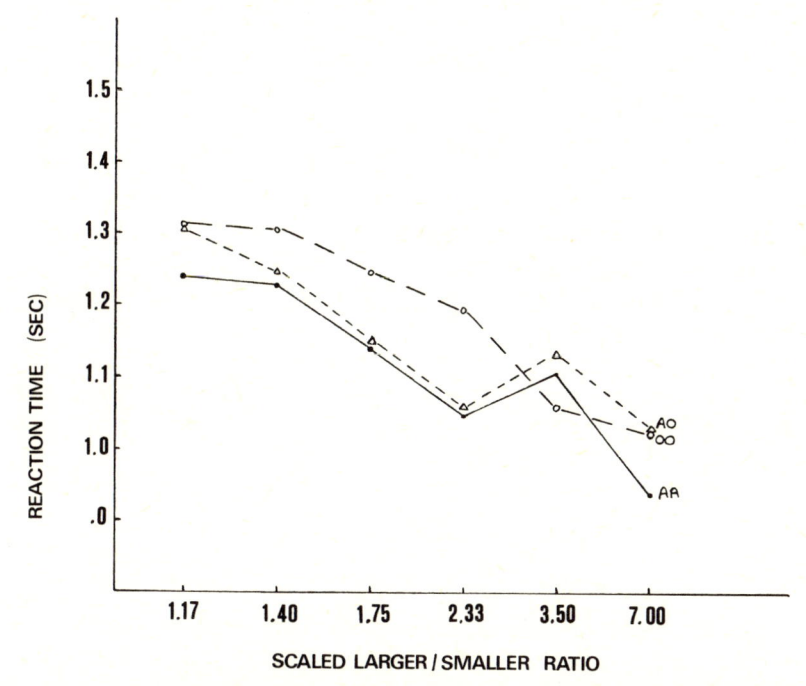

Fig. 3. Mean reaction time for choosing larger member of animal—animal (AA), animal—object (AO) and object—object (OO) pairs as a function of relative size differences (ratios based on scaled size according to average group ratings on a 9-point rating scale).

network would be extended in some fashion to capture the idea that *bear* is larger than *wolf* is larger than *duck*, and so on. Anderson and Bower do not have an example of a representation of multiple size differences, but they do have an analogous one involving relative position information (p. 408). Relative size differences presumably would be represented in a similar format, as shown in Figure 4.

There are at least two problems with such a representation. The first is that the network would have to be extremely large in order to accommodate all the relative size information that we have stored in our knowledge of the world. The second and more important problem is that the model cannot explain the function relating reaction time to size differences unless an analog process is assumed at some point in the system. That is, size difference must get represented in the propositional structure

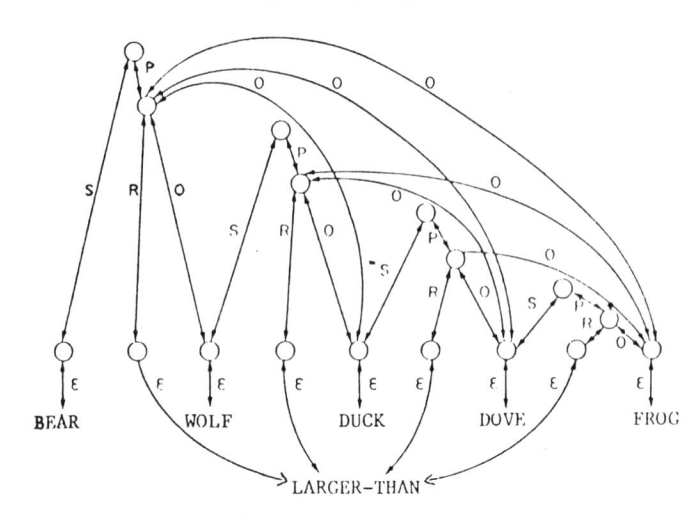

Fig. 4. How HAM might represent relative size differences (based on Figure 13.5 in Anderson and Bower, 1973, p. 408).

so that the smaller the difference for a given comparison, the farther apart the concepts would be in a functional sense. Nothing like that is included in the model. This contrasts with the analog or image theory, which predicts the function directly on the basis of the assumption that comparisons in such an internal system are essentially equivalent to perceptual comparisons.

We went on to obtain more direct evidence that the analog processes indeed involve the visual memory system. First, we extended the comparison task to pictures. That is, instead of pairs of animals or object names, we presented pairs of line drawings of the same items. If the size comparison involves a visual memory system, the reaction time should be faster with the pictures than the words because the picture has more direct access to the visual image system than do words. To test this, we compared a relatively constant size difference, according to our norms, sampling animal—animal, animal—object, and object—object pairs. The results were precisely as predicted. The reaction times were on the average about 170 msec faster for pictures than words. I will show the actual data in connection with the next phase of the experiment.

The design feature in the further extension involved the following task. Some of the picture pairs depicted a congruent size relationship like that shown in Figure 5A, where the zebra is physically larger than the lamp. The

Fig. 5. Examples of congruent (A) and incongruent (B) size relations between pictured items in the physical size–memory size conflict study.

size relationship was incongruent in other pairs, as in Figure 5B, where the zebra is now pictured as smaller than the lamp. Still other pictured pairs were equal in size. Precisely the same thing was done with printed words, that is, the pairs of words were presented so that one was in small print and the other large, as in Figure 6. The clear prediction here was that, if the size comparisons involve the visual system, there should be a conflict between the tendencies to respond to the larger object in the picture and the larger one in memory when the pictured size relation is incongruent with the conceptual size relation. Thus the reaction time should be slower for the incongruous than the congruous picture condition. This effect should not occur for words differing in printed size, or at least the effect should be

A

LAMP   ZEBRA

B

LAMP   ZEBRA

Fig. 6. Examples of congruent (A) and incongruent (B) size relations in the word condition of the physical size—memory size conflict study.

smaller, because the words must first be read before the images can be aroused and compared. The results are shown in Table I and Figure 7. As predicted, the reaction times were slower for the incongruent than for the congruent condition, and this difference was significantly larger for pictures than for words. Note that the results also indicate that the reaction time was generally faster for pictures, as I already mentioned. These results are completely consistent with an imagery interpretation of the size comparisons involving long term memory knowledge. Moreover, the picture—word differences support the dual coding idea that the image system is accessed more directly with pictures than with words as stimuli.

TABLE I

Mean RT (seconds) for choosing conceptually larger object as a function of relative sizes of picture and word pairs (S = small, L = large physical size)

	Stimulus pattern size			
	SS	LL	SL(LS) Congruent	SL(LS) Incongruent
Pictures	0.675	0.647	0.596	0.694
Words	0.859	0.837	0.803	0.800

IMAGES, PROPOSITIONS, AND KNOWLEDGE 69

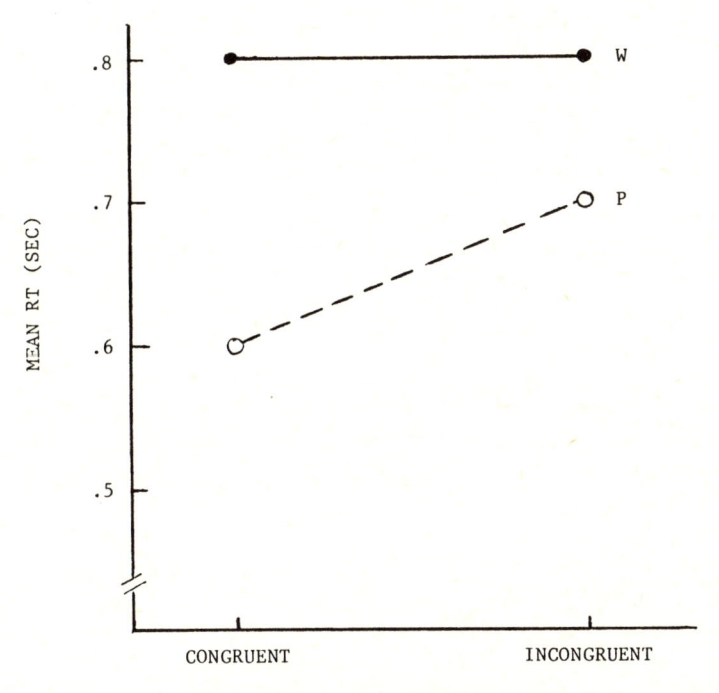

Fig. 7. Reaction time for choosing the conceptually larger of two objects presented as pictures (P) or as words (W) when physical size and conceptual (memory) size differences are congruent and when they are incongruent.

These results would not be predicted by any of the abstract semantic feature or propositional theories that I am familiar with, nor do I think that they can be explained by them after the fact. However, since others are likely to try to do so, let me summarize one final extension that was predicted from the imagery model and must be explained by any serious alternative. The extension is relatively trivial by itself, but not in contrast with the evidence of conflict between perceptual and long term memory information. I was looking for a way to reverse the conflict effect using the same stimuli and presumably the same underlying mechanism. It turned out to be easy: simply ask subjects to say which object appears to be farther away. Since the pictures contain no perspective or other distance cues, the response would have to be based entirely on relative size. Now the

TABLE II
Mean reaction times for relative distance and conceptual size comparisons of pictured object pairs

	Pictured size			
	SS	LL	LS Congruent	LS Incongruent
Distance	1.659	1.818	2.296	1.409
Size	0.675	0.647	0.596	0.694

previously incongruent conditions should yield the fastest decision time because the conceptually larger object, being pictures smaller than its companion, should clearly look farther away. Conversely, it should be most difficult to say which object is farther when the relative sizes of the pictured objects are congruent with their normal memory size. The results are shown in Table II together with the size comparison data. You can see that the prediction was confirmed. The pattern of reaction times for distance judgments was exactly the reverse of the pattern for size comparisons.

The general significance of these results is that the kind of memory psychophysics that is involved in the size and distance comparison tasks can be extended to comparisons on other physical dimensions in the perceptual world. For example, when I ask people "Which is darker, a lime or a cucumber"? they can tell me, and they also say that they infer the answer from memory images. The imagery model predicts that such a decision will be made more quickly for named objects that differ greatly in brightness than for ones that usually differ only slightly on this dimension. We are planning such extensions as well as others designed to map out the various dimensions of perceptual memory information that must be included in our knowledge of the world.

7. GENERAL CONCLUSIONS

In conclusion, I have argued that two distinct psychological systems are needed to represent the differences between language and knowledge of the world and that there is a functional continuity between perception, memory knowledge, and imagery. More generally, I am arguing for a psychological approach to mental representations rather than metaphors founded on our technology for recording information. The mind is not a wax tablet, or a tape recorder, or a computer. It does not contain pictures, words, or propositions. Such metaphors make it easier for us to talk about mental events, but they are otherwise limited in their usefulness and potentially misleading. Their limitations should not be imposed on the organism. The criteria for a psychological model should be what the mind can do, so why not begin with a psychological metaphor in which we try to extend our present knowledge about perception and behavior to the inner world of memory and thought? I have presented evidence that visual imagery is rather like active visual perception, but it is also different in a variety of ways that we can guess at from our subjective experiences although we have hardly begun to study the differences objectively. The approach of comparing mentally imagery with active perception is a far cry from comparing images to static pictures or abstract propositions, although the difference has not always been recognized. Pictures and propositions tend to divert attention away from the psychological phenomena to the properties of photographs and logical machines. Mind becomes an imitation of the model. The perceptual metaphor instead holds the mirror up to nature and makes human competence itself the model of mind.

GEORGE W. BAYLOR AND BERNARD RACINE

MENTAL IMAGERY AND THE PROBLEMS OF COGNITIVE REPRESENTATION: A COMPUTER SIMULATION APPROACH*

Protocol analysis and computer simulation are analytic tools that can be used to try to understand the role of visual mental imagery in the human problem-solving process. This approach, a theoretical psychology that draws heavily on artificial intelligence and experimental information-processing psychology, tries to account for stretches of human behavior by constructing computer programs that do the same thing in the same way that people do. To the extent that the running computer program accomplishes the task at hand, the program, which is a detailed and explicit formulation of mental structures and processes, provides a plausible model of how such knowledge can be represented and organized. Of course, as one moves towards psychologically more sophisticated models, more stringent process criteria for testing the model are imposed.

The work presented here is theoretical and empirical but it is not experimental. Thus, in de Groot's (1969) spiraling cycles of scientific endeavor, going from theory formation to theory or hypothesis testing, this work falls squarely on the side of theory formation. That does not mean it is not amenable to experimental testing — quite the contrary — only that the study presented here does not get that far.

Specifically, this work deals with what Guilford (1967) in his factor analytic language has called cognitive figural transformations (CFT), that is, with how people transform figural or imagerial information in their heads. Many years ago, Guilford, Fruchter and Zimmerman (1952) isolated CFT by constructing two spatial visualization tests: Spatial Visualization I, a paper folding and cutting test, and Spatial Visualization II, a block painting and dicing test. For the most part, the information about the problems is presented verbally so that the subject must construct a visual mental image of the situation; in order to obtain the correct information about the final state he must follow through in his own thinking how the transformations affect the image. In the same vein Moran (1973a, b) designed a 'path task' that requires the subject to synthesize a mental image of a sequence of compass instructions.

Within the confines of these three tasks, the plan of this paper is (1) to

John M. Nicholas (ed.), Images, Perception, and Knowledge, 73–93.
All Rights Reserved.
Copyright © 1977 by D. Reidel Publishing Company, Dordrecht-Holland.

present some protocol data on the paper folding and cutting test; (2) to analyze these data; (3) to present a program that generates certain aspects of this protocol; (4) to look at some plausible representational systems for visual mental imagery that (5) make realistic assumptions about the nature of the underlying information-processing mechanism.

1. THE PAPER FOLDING AND CUTTING TEST

1.1. The Task Instructions

First, the subject, a male graduate student in psychology[1], was instructed to imagine a square piece of paper that he would be asked to fold mentally several times; he was then told that on the resulting folded paper, he would be asked to imagine certain cuts. In his mind S must then unfold the paper and indicate what the unfolded paper with its cuts will look like.

Second, S was given the usual thinking aloud instructions — that he try to say everthing that goes through his head — as well as special 'luminous finger' instructions: in a darkened room S had a small flashlight attached to his finger. With his illuminated finger he was asked to trace the sequence of images that passed through his head. Both the verbal and graphic traces of his behavior were video-tape recorded.

Third, the experimenter, E, then presented one of the problems:

Imagine that you have a square piece of paper 2 inches by 2 inches. Take the left side and fold it onto the right side. Take the base of the figure resulting from the first fold and fold it onto the top. Okay? Now, in your own words repeat back to me how you are to fold the paper. (Racine, 1971, p. 56, our translation)

Finally, E showed S the following sketches, the first two of which depict the folds that had just been described verbally. The third sketch contained the new information as to how the folded paper was cut. In this problem four small triangles were snipped off the four corners of the paper. E removed the sketches from S after a few seconds and asked him to solve the problem outloud while tracing everything with his finger.

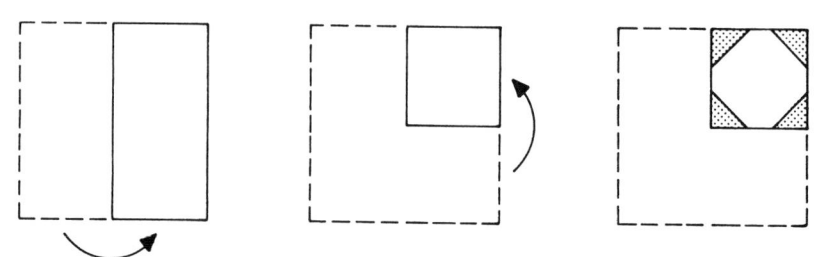

1.2. The Data

For the above problem the data consist of two parallel traces of S's mental activity: a verbal and graphic protocol that lasts a little over three minutes (204 s). The protocol falls naturally into three phases: an encoding phase in which S constructs a mental image of the paper with its folds and cuts as communicated by E; a problem-solving phase in which S indicates both verbally and graphically how he unfolds the 1 inch by 1 inch paper with its cuts into a 2 inch by 2 inch paper with its configuration of cuts; and a checking or verification phase.

Figures 1A and 1B are illustrative of how the protocol is transcribed from the video-tape. Figure 1A presents the beginning of the problem-solving phase: on the left-hand side S says (at 8) how he unfolds the 1 inch by 1 inch paper in 'a rectangle 2 inches by 1 inch by 2 inches by 1 inch'. Concomitantly, on the right-hand side of Figure 1A, the subject traces this unfolding process with his finger, starting at Ⓢ and ending at Ⓔ. In transcribing the graphic protocol a grid was placed over the television screen so that the absolute position and relative sizes of S's finger sketches, in inches, are preserved on the graphic protocol. The left-right orientation is reversed since the subject was filmed from in front.

At 9 and 10 S traces the orientations of the triangles in the 'upper left corner' and 'in the upper right corner', respectively. Then at 11 he unfolds the paper in order, at 12 and 13, to 'find in the same fashion, uh, a triangle like this in the lower left corner and a triangle like this in the lower right corner'.

Figure 1B is the tail end of the problem-solving phase once S gets to working out the consequences of the right to left fold on the center cut. As he says (States 31–33): "In re-unfolding the left side towards the right side (*sic*) what I had here, like this, will give a diamond in the center." Scrutiny of the graphic protocol reveals his imagerial representation and transformations of the center cut until it becomes the 'diamond in the center'.

Racine (1971) analyzed all 204 s of this protocol, comparing, in particular, the frequency and diversity of information found in the two parallel traces. Information on the planning of actions, for example, appears only in the verbal protocol, which, in retrospect, is not surprising since this kind of information finds no simple expression in the graphic mode. For the rest, however, the graphic protocol offers three additional types of information. First, the precise orientation of the cuts appears only in the graphic trace though they are referred to imprecisely in the verbal protocol. Of course, had the subject taken the time, presumably he could have

76   GEORGE W. BAYLOR AND BERNARD RACINE

State	Verbal Protocol	Graphic Protocol
	8. Now, I'm going to unfold ... humm ... Well, I unfold the part above, I bring it down below which is going to give me a rectangle 2 inches by 1 inch by 2 inches by 1 inch.	
	9. Now, in the part up above, I'll always have my triangle like this ... humm ... upper left corner.	
	10. In the upper right corner, I'll have the same kind of triangle oriented in the same way.	

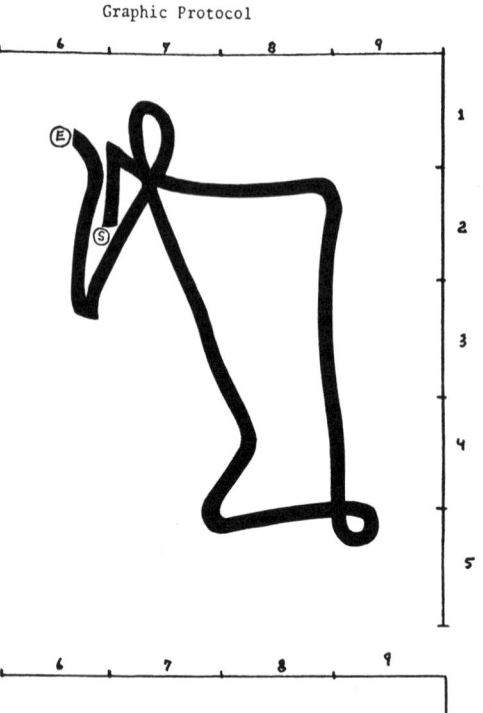

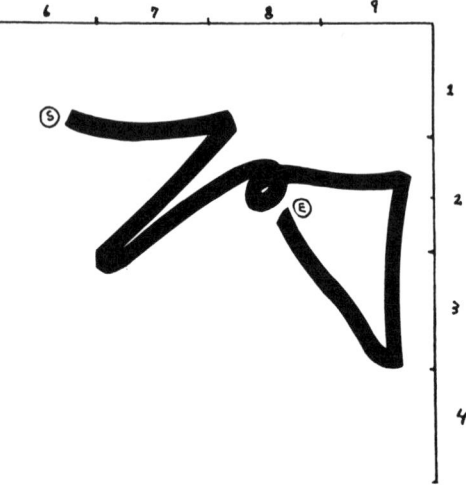

State	Verbal Protocol	Graphic Protocol

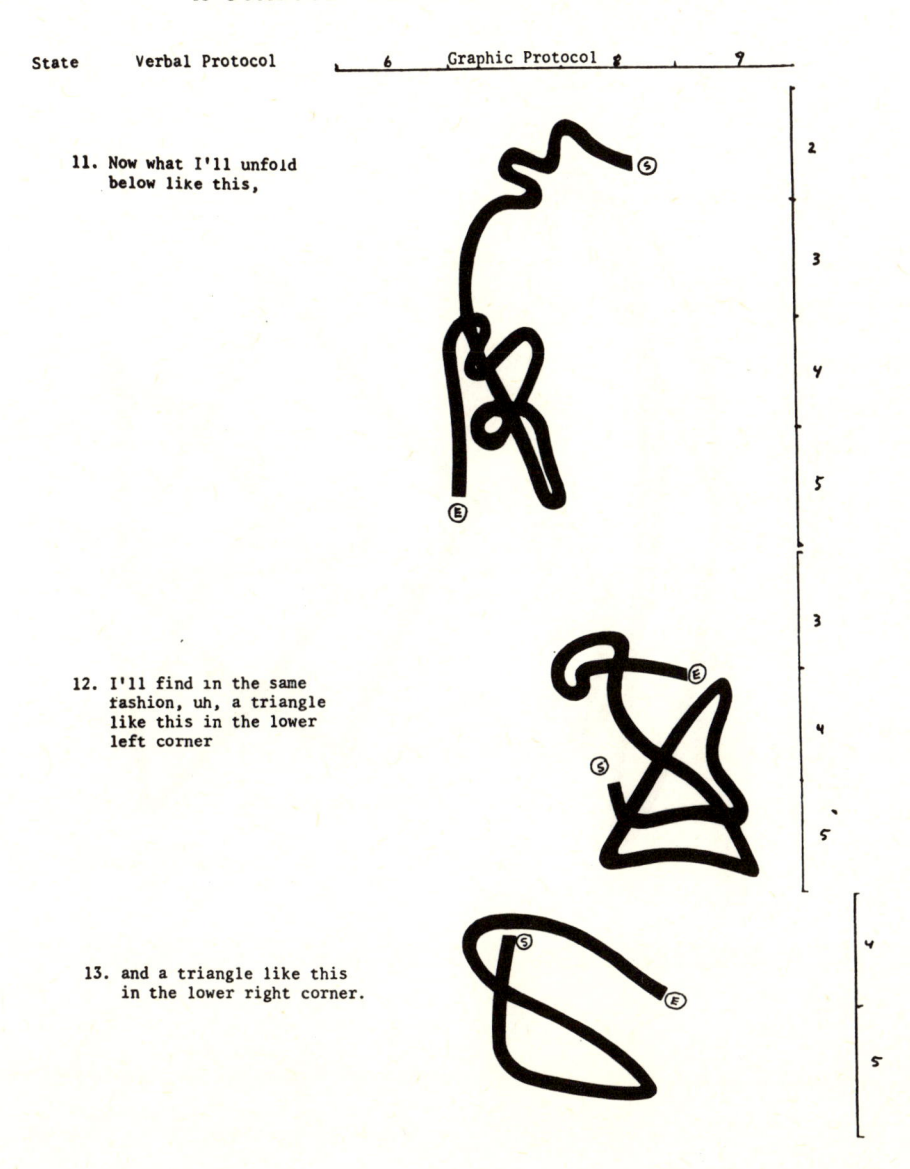

11. Now what I'll unfold below like this,

12. I'll find in the same fashion, uh, a triangle like this in the lower left corner

13. and a triangle like this in the lower right corner.

Fig. 1A. Verbal and graphic protocol for Subject CS for Episode III (States 8–13).

78   GEORGE W. BAYLOR AND BERNARD RACINE

State   Verbal Protocol   Graphic Protocol

State 31. In re-unfolding the left
          side towards the right
          side (sic)

State 32. what I had here, like this,

State 33. will give a diamond in the
          center.

Fig. 1B. Verbal and graphic protocol for Subject CS for Episode VII (States 31–33).

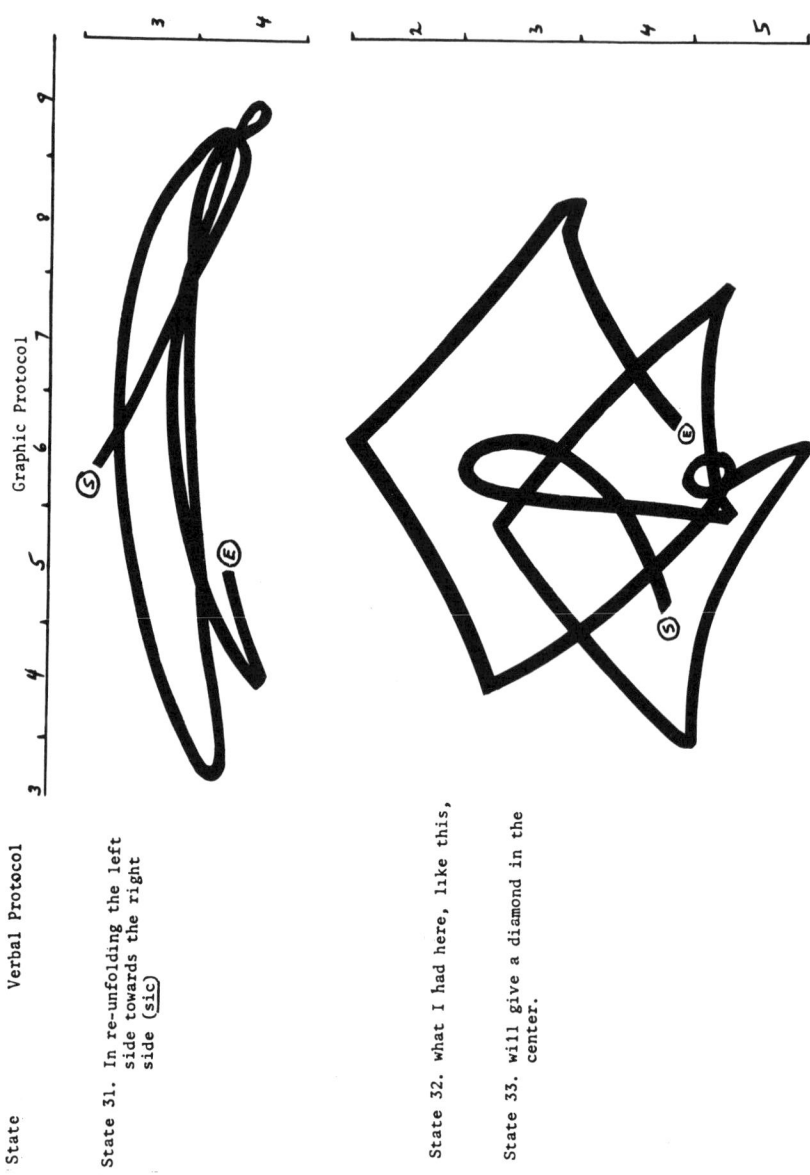

furnished precise, if cumbersome, verbal descriptions. More importantly, second, the graphic protocol reveals the surprising extent to which S maintains an isomorphism between the size and location of his mental paper and what a real physical paper would look like. On the other hand, third, there is a zoom phenomenon: once S centers his attention on a particular region, he enlarges the surface of the paper on which he is working. There is also a zoom in the other direction when S works on the paper as a whole.

These observations from the protocol of one subject on one problem must not be over-generalized. Still, they are indicative of the relative strengths and weaknesses, of the expressive powers, of two distinct 'languages' for externalizing mental processes.

1.3. The Problem Behavior Graph (PBG)

Once such a protocol has been transcribed *in toto*, it is possible to code it as a problem behavior graph or PBG, a data analysis technique largely developed by Newell (1968) and Newell and Simon (1972). Such a coding views the problem solver as moving through states of knowledge by means of operators that transform one state into another.

Thus, for the entire paper folding protocol, Subject CS is viewed as moving through 43 states of knowledge by means of five operators. The first two operators, FOLD and CUT, occur in the initial encoding phase of the protocol and correspond, respectively, to S's mental operations of folding the paper from left to right or from top to bottom and of cutting out the small triangles. In the problem-solving phase the principal operator is UNFOLD, either from top to bottom ($UNFOLD_{T-B}$) or from right to left ($UNFOLD_{R-L}$). There is also an operator called FOCUS, which homes in on the particular cut that is assumed to be the current content of the subject's mind's eye. Finally, there is an operator called FUSE, which chunks two cuts together to form a new figure. FUSE is meant to emphasize the possibility of the physical proximity of two figures in the mind's eye giving rise to a new figure.

Figure 2 presents the PBG for Subject CS. It is to be read from left to right and then down. The 43 graphic states of knowledge are numbered and organized into eight episodes: Episodes I and II representing the encoding phase, of the folds and cuts, respectively. The problem-solving phase runs from Episodes III through VII with each episode initiated by an UNFOLD operator whose effects on certain cuts are then followed through. There should be an Episode VIII that belongs to the problem-solving phase but S

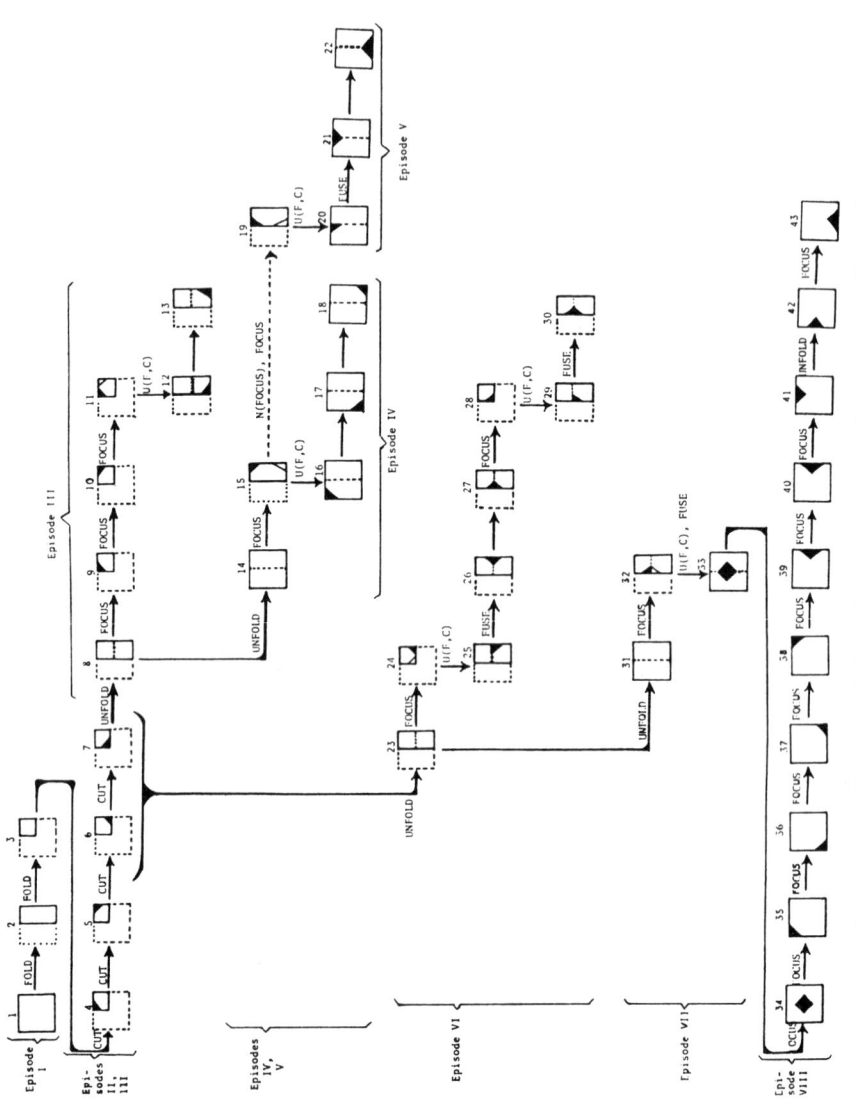

Fig. 2. Problem behavior graph for Subject CS on a paper folding and cutting problem.

omits it; instead Episode VIII is the verification phase where, in fact, S catches his error of omission and so turns in a successful problem solution.

The PBG traces S's sequence of actions. Once he FOLDs the paper into a 1 inch by 1 inch square in Episode I and CUTs off the four corners in the order indicated (States 4, 5, 6, and 7) in Episode II, he decides to 'unfold the part above . . . bring it down below' at State 8. He then FOCUSes on the triangle in the 'upper left corner' at State 9 and the 'upper right corner' at State 10. These are the same two cuts, in the same order, that he encoded at States 4 and 5, respectively. At State 11 the 'what I'll unfold' of the subject is coded on the PBG as: ◨ . The darkened triangle refers to the fact that S FOCUSes explicitly on the first cut (F) while apparently maintaining its complement (C) in short-term memory (STM) — the undarkened triangle. He then applies the UNFOLD operator, abbreviated as $U(F,C)$ on Figure 2, first on the upper left-hand triangle to yield State 12 and then on the upper right-hand triangle lurking in STM to yield State 13.

Episodes IV and V show S working out the consequences of unfolding the 1 inch by 2 inch rectangle from right to left for the four cuts of Episode III. Once he FOCUSes at State 15 on the upper right-hand triangle of the 1 inch by 2 inch rectangle and its complement in the lower right-hand corner, the unfolding operation yields States 16, 17, and 18. The 'next focus' — $N(FOCUS)$ — at State 19 is the original upper left-hand triangle with its short-term memory complement. Note at State 21 that S FUSEs the results of States 19 and 20, giving rise to a large triangle up above. FUSE is a name for this process, which appears to be a kind of image operation based on physical proximity. Immediately thereafter S arrives at State 22 by a kind of reasoning by symmetry: 'And *idem* below' as S puts it.

In Episode VI, S returns to the other two original cuts that he encoded way back in States 6 and 7. He unfolds the 1 inch by 1 inch paper (State 23) and then traces what happens to the two little triangles. At State 26 he again FUSEs what he has just unfolded and notes the symmetry (State 27). He then checks State 27 with the explicit image operators to make sure he got it right (States 28—30).

Figure 1B, the protocol segment for Episode VII of Figure 2, was discussed earlier. At State 31 S unfolds the paper from right to left; traces with his finger at State 32 the upper part of the internal triangle with its assumed complement in STM, which was carried along from State 30; and unfolding the paper at State 33, he fuses the two parts to arrive at 'a diamond in the center'.

There follows the verification phase, Episode VIII, where at State 42 S notes his error of omission and so successfully completes the problem.

1.4. The Program

To understand further the nature of visual mental imagery in the problem-solving process, the next step is to put together a plausible representation of the subject's mental images and to program the operators that make use of this representation. The execution of the operators according to some organizational scheme or strategy then generates the actual sequence of knowledge states the subject went through, or some approximation thereof. Obviously, the closer the approximation, the better the simulation model.

While the protocol only provides a few bench marks as to the nature of the representational system, it is more revealing as to the organizational structure of the principal problem-solving operators, UNFOLD and FOCUS. They appear to be organized into a heuristic search method known as a depth-first strategy (see Newell and Simon, 1972, and Newell, 1973a, for further discussion), which provides for back-up to previous states as part of the control mechanism. From the psychological point of view the use of such a strategy implies that the subject is able to remember these states in some intermediate or long-term memory; otherwise, he would not be able to return to them later on. In Figure 2 these returns to earlier states are, in fact, the episode boundaries for the problem-solving phase of the protocol.

A program that generates the S's problem-solving phase, omitting the FUSE operator, can be written as follows:

$\langle$ UNFOLD$_{T-B}$, FOCUS, N(FOCUS) $\{$UNFOLD$_{R-L}$, FOCUS, N(FOCUS)$\}\rangle$

To interpret this set of instructions requires another simple program, an interpreter, that reads from left to right, executing the operators as it proceeds to the limit of the embedding whereupon it cycles within the embedding until the 'alphabet' of cuts is exhausted, that is, N(FOCUS) → NIL. Thus, the choice of cuts is also determined by the program. The instruction, N(FOCUS), means that the next pair of cuts on the list is to be prepared for later. Next, the interpreter backs up, moving from right to left until a non-exhausted generator is found, whereupon it proceeds from left to right again.

Figure 3 presents the interpreted program that generates most of the states of knowledge of the problem-solving phase of the protocol. The graphic representations of the generated states appear in Figure 2. The last three instructions in Figure 3, enclosed in a box, generate hypothetical

UNFOLD$_{T-B}$ — — > State 8
  FOCUS (States 9–11) — — > States 12, 13
  N(FOCUS) — — > (State 24)
    UNFOLD$_{R-L}$ — — > State 14
      FOCUS(STATE 15) — — > States 16–18
      N(FOCUS) — — > (State 19)
      FOCUS(State 19) — — > States 20, 22
      N(FOCUS) — — > NIL
UNFOLD$_{T-B}$ — — > State 23
  FOCUS(State 24, 28) — — > States 25, 29
  N(FOCUS) — — > NIL
    UNFOLD$_{R-L}$ — — > State 31
      FOCUS(State 32) — — > State 33

N(FOCUS) — — > State X corresponding to ▢

FOCUS(State X) — — > State Y corresponding to ▢

N(FOCUS) — — > NIL

Fig. 3. Interpreted program that generates the problem-solving phase of the PBG for Subject CS on a paper folding and cutting problem.

States X and Y. This is meant to show how S could have picked up immediately the triangular cut in the center of the left-hand side of the 2 inch by 2 inch paper if he had just followed 'his program' through to completion!

1.5. Discussion

This is as far as the analysis of the paper folding and cutting test goes. The methodology has permitted certain insights into the nature of the image space and the organization of its operators, but the resulting model is still rudimentary. As a simulation model, it fails to account for the FUSE operator (at States 21, 26, 30, and 33) and for the symmetry operation (at State 27). Moreover, it does not offer any explanation as to why the subject does not unfold the 1 inch by 2 inch paper from right to left for a final Episode VIII in the problem-solving phase. Finally, the model ignores the initial encoding and final verification phases.

Two major incompletenesses are more serious. First, no representational system for the mental images was actually specified. Thus, while a program for the organization of the operators was proposed, the actual input/output specifications of the operators and the content of the states of knowledge have not been instantiated in a running computer program. This is a serious omission since it is part of the philosophy of the computer simulation approach that much, if not all, of one's theory is embedded in the running program. Explicit representational systems are presented for the two other tasks below.

Second, the analysis contains no explicit model of short-term memory. It has already been noted that in executing the above program the interpreter requires a stack so that it can keep track of its place in the execution of the instruction set. This is one form of relatively short term memory. The FOCUS operator implicitly makes use of a STM too since S always appears to carry along *two* cuts, the focus and its complement. While the consequences of the first cut are computed across the unfolding transformation, the second must not be forgotten. The model of the path task presented below contains an explicit STM.

2. THE BLOCK VISUALIZATION TEST (BVT)

Baylor's work on Guilford's block visualization test (BVT) has been discussed in detail elsewhere (Baylor, 1971, 1972, 1973). He carried out an analysis very similar to the one presented above of the verbal protocol of a single subject on a number of BVT problems though he did not have the benefit of a parallel graphic protocol as Racine (1971) had not yet introduced the 'luminous finger' technique. The computer program modelled rather closely the behavior of the subject on several questions, but what is of interest here is just how the system of mental imagery was represented in the computer.

First, try solving one of the BVT problems mentally, without the aid of paper and pencil:

> The top bottom and one side of a 3 inch cube are painted blue and the block is then cut into twenty-seven 1 inch cubes.
> How many cubes are blue on three faces?
> How many cubes are blue on one face?
> How many cubes have no blue faces?

Most subjects experience visual mental imagery on this task: they initially color the three faces of the 3 inch cube, compute or accept the feasibility of 27 one inch cubes, and then start worrying about what this means when they are confronted with the first question. They then start problem solving on the mental image they have constructed. How was this simulated?

2.1. *The Representational System*

A cube or block is composed of six faces, twelve edges, and eight vertices related to each other in the three dimensions of the mind's eye: UP-DOWN, RIGHT-LEFT, and TOWARDS-AWAY. Thus, the image of a block or cube in general is represented as a list of eight vertices, (V1 V2 V3 V4 V5 V6 V7 V8), connected to its component faces, edges, and vertices in a one-way directed graph structure made up of 48 pathways. It was observed, moreover, that the subject always had a right side projection of this base block in mind such that V4, e.g., is away from V2, to the right of V3, and up from V8:

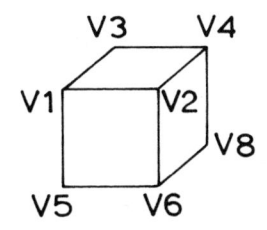

```
         V3      V4
   V1 ┌──────┐
      │   V2 │
      │      │ V8
      └──────┘
      V5     V6
```

In terms of the number of interconnections in the underlying graph structure, there is a difference between such an image representation, 'seen' from a single point of view, and a richer perceptual one, which could well be made up of multiple copies, one for each desired point of view or block rotation.

Now given that a person has such a general description of a block stored in memory, he must still be able to construct a specific representation when confronted with the above sort of problem. That is, he must be able to particularize the block into a three inch cube with three blue faces that is sliceable into 27 one inch cubes. In the computer such descriptive information is appended to the definition of the base block by means of property lists. Thus, the base block, (V1 V2 V3 V4 V5 V6 V7 V8), whose TOP is (V1 V2 V3 V4), also has a DEPTH, WIDTH, and HEIGHT of 3, and a TOP, FRONT, and BOTTOM colored BLUE. Cutting up the block into 27 little cubes means making HORIZONTAL, LEFT-RIGHT-VERTICAL, and BACK-FRONT-VERTICAL SLICES. The following list structures show how some of this information is represented:

(V1 V2 V3 V4 V5 V6 V7 V8)
| TOP: (V1 V2 V3 V4)
| BOTTOM: (V5 V6 V7 V8)
| LEFTSIDE: (V1 V3 V5 V7)
| RIGHTSIDE: (V2 V4 V6 V8)
| FRONT: (V1 V2 V5 V6)
| BACK: (V3 V4 V7 V8)
| DEPTH: 3
| WIDTH: 3
| HEIGHT: 3
| THREE-D: (3 3 3)
| HAVE: ((TOP FRONT BOTTOM) COLOR BLUE)
| PARTS: ((CUBES THREE-D (1 1 1)) HOW-MANY 27)
| SLICES: (HORIZONTAL LEFT-RIGHT-VERTICAL BACK-FRONT-VERTICAL)
| NAME: BLK

where the image of each of the block's faces (and edges), for example, the blue top, is represented as follows:

(V1 V2 V3 V4)
| FRONTEDGE: (V1 V2)
| BACKEDGE: (V3 V4)
| LEFTEDGE: (V1 V3)
| RIGHTEDGE: (V2 V4)
| DEPTH: 3
| WIDTH: 3
| COLOR: BLUE
| NAME: TOP

The effect of slicing the block, once the number and orientation of the slices has been figured out, is to create new vertices, in fact, four new vertices for each slice. This in turn leads to the creation of new edges, faces, and pieces, some of which may possess the sought-after properties required by the BVT questions. On this problem the program creates nine new pieces, which are independent of each other in the sense that their intersections, necessary to create 27 one inch cubes, are not at first computed. Moreover, these nine pieces are only partial images in that only their top, front, and bottom faces are defined.

The stage is now set to focus on the block's blue top (and front) in order to see which pieces could have three (one, or zero) blue faces. The two principal problem-solving operators, Process Block and Tally, were programmed for this purpose: to make use of the image representation in order to track the effects of the slice transformations on the distribution of blue faces on the new pieces — in the same way that in the paper folding test solving the problem consists of tracing the effects of unfolding the paper on the distribution of cuts on the larger paper. Baylor (1972) should be consulted for a detailed explanation of how the problem-solving operators bring about a solution.

The point here is that it was possible to create a representational system for mental imagery that a computer program made use of in order to solve BVT problems like the one presented above. While the model is far from perfect, it does appear to capture a number of structural and functional properties of the subject's system of mental imagery: the representation (1) of a base block in terms of interconnected vertices, edges, and faces; (2) of a particularized block that meets the specific requirements of the BVT problems; (3) of a cut up block with partially defined images of pieces; and (4) of a focus face that serves as a work space in the mind's eye. Well-defined problem-solving processes make use of these images in order to extract new information and solve BVT problems. There is at least this degree of isomorphism between the information-processing model and the imagerial phenomena it is purported to explain.

3. THE PATH TASK

In Moran's (1973a, b) path task E presents a sequence of directions, one at a time, to the subject ('north', 'south', 'east', or 'west'). For each direction S is to imagine a line segment of unit length and to describe the path as he constructs it. His overall objective is to understand, remember, and talk about the global path configuration. He could also be requested to answer questions about the final configuration.

Problem 1, which the reader should try, consisted of the following twenty directions: N, E, S, E, S, E, S, E, N, E, S, S, E, E, N, E, N, W, W, N. A protocol on this task showed the subject building up the path by trying to assimilate each new direction to the path he has already constructed. Thus, the first four moves in the above problem were encoded as follows: *S* starts with 'a vertical line'; then 'an angle going up and to the right'· 'three sides of a box: everything but the bottom is closed'; 'a box sticking up and a line'. By Move 7 the pattern was 'a vertical line and essentially three steps going east and down' (Moran, 1973a, pp. 343–344). In trying to define such properties of an image space as closure, for example, it is important to note how many subjects are cognizant between Moves 11 and 19 of what Moran's subject calls 'this two by two box that you never quite ever finished'.

3.1. *The Representational System*

Now Moran constructed a computer program that modelled in fine detail his subject's protocols for two of the path problems. The internal representation of the imaged paths consists of a hierarchic structure of expressions describing the recognized figure types in the paths. For example, in the above protocol fragment by the end of the ninth move, the program has constructed the following representation:

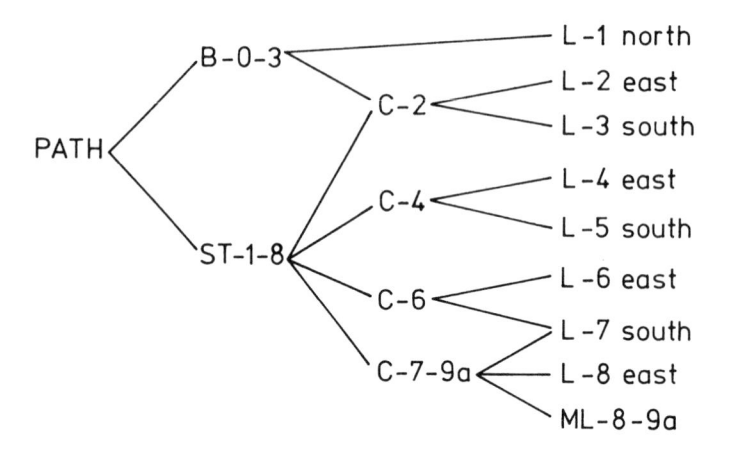

```
                    ┌─ B-0-3 ─┬──────────── L-1 north
                    │         │            ┌─ L-2 east
                    │         └─ C-2 ──────┤
                    │                      └─ L-3 south
         PATH ──────┤                      ┌─ L-4 east
                    │         ┌─ C-4 ──────┤
                    │         │            └─ L-5 south
                    └─ ST-1-8 ┤            ┌─ L-6 east
                              ├─ C-6 ──────┤
                              │            └─ L-7 south
                              └─ C-7-9a ───┬─ L-8 east
                                           └─ ML-8-9a
```

The *L*-expressions represent lines; e.g., *L*-1 is a symbolic expression of the following sort: (*L*-1 LINE VERT SOUTH *P*-0 NORTH *P*-1 MOVE NORTH) where the *P*-expressions anchor each end point of the line. The program creates such complex figures as boxes, corners, crenelations, lines, points, *S*-figures, *T*-figures, steps, and areas. *C*-2, a corner, is thus described as: (*C*-2 CORNER *A*-1-3 *P*-2 NORTH *L*-2 EAST *L*-3 MOVE *L*-2 *L*-3 STOP) meaning that the corner in question consists of an area delimited by edges L-2 and L-3, which meet at vertex P-2, and was generated in the order L-2, L-3.

The C-expressions are corners made up of their respective lines. C-7-9a is particularly interesting because S expected another corner at this point in the protocol and thus noted a 'missing line' (ML-8-9a) that should have gone from points 8 to 9. After Move 8 S says· "You're continuing the same pattern . . . you're beginning the fourth step." When his expectation is not confirmed at Move 9 he says· "Oh shit. Now you've, uh, enclosed a box on three sides."

The step pattern, ST-1-8, beginning at point 1 and ending at point 8, is defined as follows: (ST-1-8 STEPS EAST SOUTH ONE C-2 TWO C-4 THREE C-6 OPEN FOUR C-7-9a END LENGTH THREE MOVE C-2 C-4 C-6 C-8-9a STOP). It is composed of the four corner expressions as well as certain additional information.

Finally, the box, B-0-3, is made up of the first vertical line, L-1, plus the corner, C-2. The program, like the subject, also constructs a corner, C-1, made up of L-1 and L-2, but is is assimilated by the box figure and so disappears from the final figure hierarchy.

3.2. Assumptions About the Underlying Information-Processing Mechanism

The program functions by means of a relatively large set (175) of condition → action rules called productions. This is permanent knowledge, stored in LTM, that S brings to the task and enables him to construct images of the sort described above. S's momentary knowledge about his current place in the problem is held in an explicit short-term memory (STM). As new knowledge is produced, new expressions are inserted in STM and old expressions are deleted, thus keeping the size (in chunks) of STM constant. The program is STM-driven in the sense that the current contents of STM determine which rule will be evoked from the collection of productions in LTM.

Without being able to do justice to the program, a few examples of the kinds of rules stored in LTM may illustrate the program's psychological foundations: The program has recognition rules that are always trying to chunk the most recent information with other expressions still in STM in order to create higher-level figure expressions. For example, the corner recognizing rule looks for two line segments that have a common end point and different slopes. There is also a recognition rule for corner expressions sharing a common leg: the expression created describes a box with three sides and open on the bottom. This accounts for how the box expression,

B-0-3, is created after Move 3. An interesting psychological feature is the creation of ML-3-0 to represent the missing side of the box, having anticipated the next move by putting EXPECT WEST in the B-0-3 expression in STM. There are also rules for monitoring expectations and comparing them with the new input: if the next move conforms to expectations the program chunks it easily into its current path configuration; otherwise, it constructs a new corner and/or line. Thus, on Move 9, both program and subject expected the step pattern, ST-1-8, to be continued but recovered when the expectation was not confirmed by creating a new line to the north and reintegrating it as part of 'an upside-down box'.

Like Newell's (1972, 1973b) PSG system, Moran's (1973a) short-term memory structure plays a crucial role in the overall behavior of the model. Assumptions about its capacity determine subsequent behavior. Moran set his STM simulations to a constant size of 20, which seems quite high when compared with the classic 7 plus or minus 2, until one realizes that this contains not only substantive, content information but also control information as to what to do next in the task. Moran (1973a, p. 139) figured, however, that 'with several (quite reasonable) changes in the program, it could run with an STM of twelve (and perhaps even less) slots and its behavior would not be significantly different from that of the present version. Speculation about how many slots there are in STM must be balanced by a consideration of how much information can be stored in each chunk. There is a trade-off in the design of a working memory between holding information in a few large chunks or in many small chunks'. Since there is no theoretically established limit, especially for familiar concepts, on how much information can be stored in a chunk, the program's chunk size appears plausible because of the close correspondence with the subject's behavior and the fact that it works within a limited memory.

4. DISCUSSION AND CONCLUSIONS

From the foregoing analyses and models, it seems clear that visual mental imagery plays a crucial representational role in the human problem-solving process. Moreover, since it was possible to construct discrete symbol manipulating systems that solve spatial visualization problems, both what is meant by an imagerial representation and the process of mentally transforming figural information (CFT) have been defined. While in the preceding brief discussion there was little insistence on the goodness of fit or closeness of correspondence between the models and the modelled, originally, the systems were presented as simulations of human protocol data. They are still ripe for experimental testing.

As to the analog-digital or continuous-discrete nature of visual mental imagery, this work, as far as it goes, would appear to support the digital, discrete point of view. As Moran (1973a, p. 181) points out: "the normal use of the terms 'analog' and 'digital' implies a continuous-discrete distinction. An analog computer uses a continuous physical feature (e.g., a voltage) to represent a continuous variable of interest. A digital computer only uses discrete symbols..." Thus, a digital computer program is *ipso facto* digital and discrete.

There is, however, another use of the word 'analog': it is in the sense of analogy. Cooper and Shepard (1973, p. 88) in discussing their work on the mental rotation of certain geometric figures say "there very likely also exists at least some degree of structural analogy or 'first-order' isomorphism between the internal representation of an object that is rotating in space and the spatial structure of the object itself." Of course, to speak of a 'structural analogy' between the mental image and the 'object itself' is simply to evoke a relation of *resemblance* between a sign and its significate. This has to do with the problem of reference, and, according to a view summarized by Fodor (1964, p. 567) 'reference is a property of two sorts of vehicles: signs and symbols'. Signs can be related to their referents either by causal connection or by resemblance. (Linguistic) symbols, on the other hand, are both arbitrary and conventional. Thus, in the cases where the sign relation is mediated by resemblance it is by its very nature not totally arbitrary. Assuming, thus, that a structural analogy or resemblance relation does hold between the mental image and the object, this reasoning leads one to regard the mental image as a sign rather than as a symbol.

Cooper and Shepard (1973, p. 87) also propose a process analogy or 'second-order' isomorphism between mental imagery and the perceptual process: 'it would be sufficient, to justify speaking of mental rotation, if the internal process that we are calling a merely imagined rotation has a functionally important component that is uniquely shared with the internal process that takes place when the same subject is actually perceiving an externally presented rotation — *whatever* the detailed nature of this process may be at the neurophysiological level.'

So, if true, a formal model of mental images and mental imagery must conserve the first-order isomorphism between the image and various structural properties of the percept as well as the second-order isomorphism between imagery and the analogous perceptual process. Such a model must in fact be a theory not only of mental imagery but also of certain aspects of perception for to say that mental imagery resembles perception is to explain nothing unless there exists (1) a detailed model of perception (which imagery resembles) and (2) a precise characterization of the formal relations upon which the resemblance depends.

Finally, given this prelude, it does seem to us in the paper folding and cutting test that the mental representations of the little pieces of paper with corners cut off here and there, as revealed by the graphic protocol, do indeed reflect a certain first-order isomorphism. Indices like the orientation, location, and size of the cuts and pieces as well as their fusion based on physical and temporal proximity indicate some degree of isomorphism between the mental representations − as conveyed by the luminous finger − and certain aspects of a real paper. This certainly does not mean that we subscribe to a view of mental imagery as mental pictures, the so-called naive imagery hypothesis (Anderson and Bower, 1973; Pylyshyn, 1973), only that the relation of resemblance or first-order isomorphism seems more compelling than what has been captured so far in any explicit representational systems. Whether this is because images are analog in nature − in the 'continuous' sense of the word − is an open question.

With respect to the second-order isomorphism, on the other hand, the models presented consist of programs of operators with particular control structures; such programs are sequences of discrete interpretable instructions in a command language. Moreover, there is nothing in the data that suggests that this process is not well modelled as a discrete, digital, 'symbolic' one (but see Cooper, 1975).

It is just possible, of course, that the mental images are of an analog, continuous nature and the program that drives them digital and discrete. Such a hybrid model − a sign manipulating system if you will! − remains to be put together; moreover, it must be expressed in some medium − be it voltage, mathematical symbols, or an information-processing list-processing language. Operators could be programmed in a list-processing language and the mental objects represented in some continuous, analog language, but this remains to be tried. Only once the formal elements and operations of such a hybrid model were coordinated with their 'real' counterparts and predictions generated about 'real' behavioral events could such a model be taken seriously.

Université de Montréal

NOTES

* This research has been supported from a grant from le Ministère de l'Education du Québec as part of the 'Programme de formation de chercheurs et d'action concertée'. Without implicating him for the views expressed in this paper, we would like to thank Tom Moran for a helpful discussion of its contents.

[1] We would like to take this occasion to thank Mr. Claude Sarrazin who not only served as subject for this experiment but also participated fully in the subsequent data analysis.

STEVEN ROSENBERG

THE SEPARATION AND INTEGRATION OF RELATED SEMANTIC INFORMATION*

1. INTRODUCTION

Each of us, as we deal with our complex environment, is constantly bombarded by new information. We must try to understand and integrate knowledge obtained from a number of different sources. This information may be perceived visually or it may be heard; it may involve the use of language, or a pictorial encoding. Regardless of the different communication forms used, people are generally successful in integrating and understanding information from a variety of different sources.

This ability raises an interesting question. Do people represent information which occurs in different forms in a modality-dependent fashion? Do they then have to 'translate' between these forms in order to understand? Or is there at some level a single system for integrating and representing related information, regardless of source? For example, to follow the action in a movie, how do we connect the linguistic expression of actions, feelings, intentions, etc. with the visually perceived action?

There are two particular areas where this question is relevant. The first involves bilinguals — people who have two language systems for expressing thought and for understanding linguistic communications. Do they have separate semantic systems for the representation of meaning or do they have only a single underlying system? (cf. Lambert and Rawlings, 1969).

Different languages are instances of a linguistic communication mode. The most common communication and understanding system besides language is the visual one — the use of pictures, imagery, and spatial relations. Psychologists have for a long time distinguished the linguistic or verbal mode of understanding from the spatial or imagery mode of representing information. They have gone so far as to localize linguistic functioning in the left hemisphere and spatial and imagery functions in the right (Gazzaniga, 1970).

An appropriate question to ask is whether people have different systems for representing information which is perceived in a visual format as opposed to a linguistic encoding.

John M. Nicholas (ed.), Images, Perception, and Knowledge, 95–119.
All Rights Reserved.
Copyright © 1977 by D. Reidel Publishing Company, Dordrecht-Holland.

This paper will discuss an information processing theory capable of explaining certain experimental evidence. Both the theory, which exists as a working computer simulation, and the experimental findings support the hypothesis that at some level only a single semantic system is required.

Consider the following paradigm. Suppose we present a person with a sentence such as "The scared cat running from the barking dog jumps on the table." This sentence is composed of four simple 'ideas': the cat is scared; the cat runs from the dog; the cat jumps on the table; the dog is barking. These four ideas can be combined in various ways to form a variety of sentences of different complexities. Eleven sentences in all can be derived from our original four-idea sentence. Besides the four one-idea sentences, there are four two-idea sentences and three three-idea sentences.

Suppose we present subjects with half these derived sentences in an acquisition task. Later, during the recognition phase of the experiment, we present subjects with the remaining half of the derived sentences. We then ask them if they have ever seen these exact sentences before. This is, in fact, a paradigm used by Bransford and Franks (1971) to test for semantic integration of related ideas.

In a situation of this sort subjects will very often say they recognize a particular sentence even though they have never before seen it. They even accept sentences such as the original four-idea sentence which are longer than any they have seen. This has been explained by assuming that people will integrate related ideas into a single semantic structure. On seeing a new sentence, they match it against that structure rather than against representations of the individual sentence.

They decide whether they have previously seen the new sentence on the basis of the completeness of the match. Since the new sentences are composed of the same basic ideas as the old, this match will generally be rather good. As a result subjects are willing to say they have seen these new sentences. Thus people do not store the actual sentences they have seen, but they do retain an integrated representation of the sentence meanings.

This finding that people abstract and integrate the meanings of sentences in order to represent and retain this information has been demonstrated by several recent studies (Bransford, Barclay and Franks, 1972; Franks and Bransford, 1972; Bransford and Johnson, 1973; Kintsch and Monk, 1972. See, however, Singer and Rosenberg, 1973; Franks and Bransford, 1974, for recent discussions of this particular paradigm).

These results suggest that the following modification of the paradigm might shed some light on the questions we have posed. First we will present subjects with half of the sentences derived from a complex four-idea sentence. Half of these sentences will be in one language, half in another. In a subsequent recognition task subjects will see sentences that have occurred before, the other half of the derived sentences, and most importantly, translations of sentences they have previously seen. How will they handle these translations? If, on the one hand, people have separate, language-dependent systems for representing knowledge, they should be able to easily reject these translations. On the other hand, if they have a single integrated semantic system, people should frequently accept translations, since there should be an appropriate match in their single semantic representation.

These ideas form the basic paradigm used in the research we will be discussing. Two experiments were run, both based on this logic. One examined the question of visual vs. linguistic encoding, and the other looked at the semantic system of bilinguals. (For a more detailed discussion of these experiments, see Rosenberg, 1974.) We will present the experimental evidence first, and then go on to discuss an information processing model able to explain these results.

2. THE EXPERIMENTAL EVIDENCE

Four idea sets (i.e., sets of eleven derived sentences) were used in each experiment. For the French-English study, translations of these sentences were developed. For the Picture-Sentence experiment, simple pictures corresponding to each of the derived sentences were formed. Figure 1 shows the picture which corresponds to the complex sentence 'The scared cat running from the barking dog jumped on the table.'

Fig. 1.

Since the design of both experiments is similar, only the Picture-Sentence experiment will be described. For each idea set there were eleven derived sentences of various complexities, and eleven corresponding pictures. In a typical condition, three sentences of different complexities and three pictures (not translations of any of the chosen sentences) from each idea set were used in the acquisition phase. Subjects saw sentences and pictures which had been previously presented — these are called Literals. They received translations of previously presented pictures and sentences, so that combinations of ideas seen previously as sentences were now seen as pictures and vice-versa. These items are referred to as Translations. Finally, they received the semantically consistent items form the original lists which had not been previously used — these are called Recognition items.

There are several questions to ask of the data. For instance, how do people treat Translations? If they have separate semantic systems they should be able to reject Translations. If they are integrating information across modalities, they should have high false alarm rates for these items. Do people have different patterns of response when the material occurs in a form other than English? Previous research has shown how these ideas are treated when presented as English sentences. A different pattern of results with another modality would suggest modality-dependent ways of representing information.

Figure 2 shows the per cent accepted for our three categories of items (Literals, Translations, and Recognitions) for both experiments. Both studies produced the same pattern of results. Items which have been seen previously (Literals) are accepted most frequently, followed by Translations. Least frequently accepted are the semantically related new items (Recognitions). Speed of acceptance (Figure 3) shows an inverse pattern to the proportions accepted results. Looking at these later results, we note that people are frequently not able to reject Translations. They accept them quite often. This indicates they often think, for example, they have seen a picture when they actually saw a sentence.

The same pattern of results is observed with English sentences, French sentences, and Pictures, as is suggested by Figures 2 and 3. Close examination of the data reveals no asymmetries that would indicate translating from one modality to the other is an encoding strategy. Previous researchers have found increasing acceptances of English sentences as they become more complex. This finding is duplicated for the present research. There are no interactions with this effect from the type of material (French, English sentences, and pictures) or the native language of a subject. The

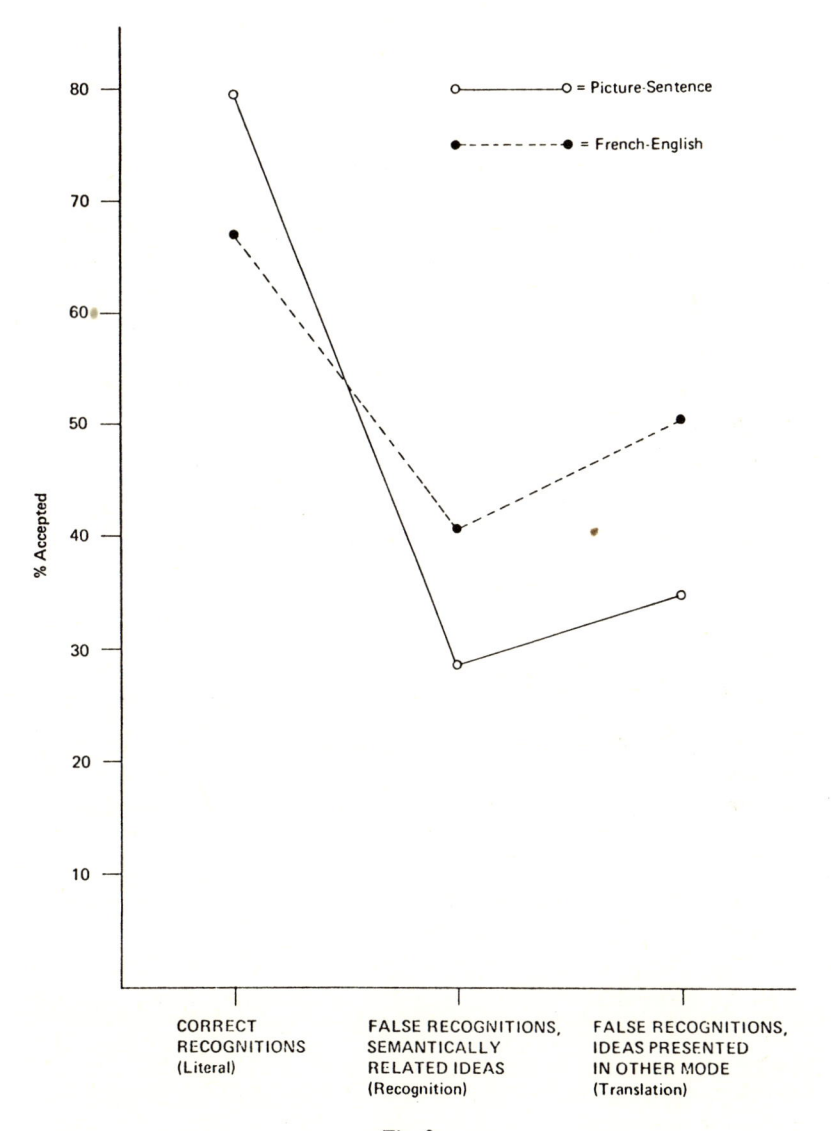

Fig. 2.

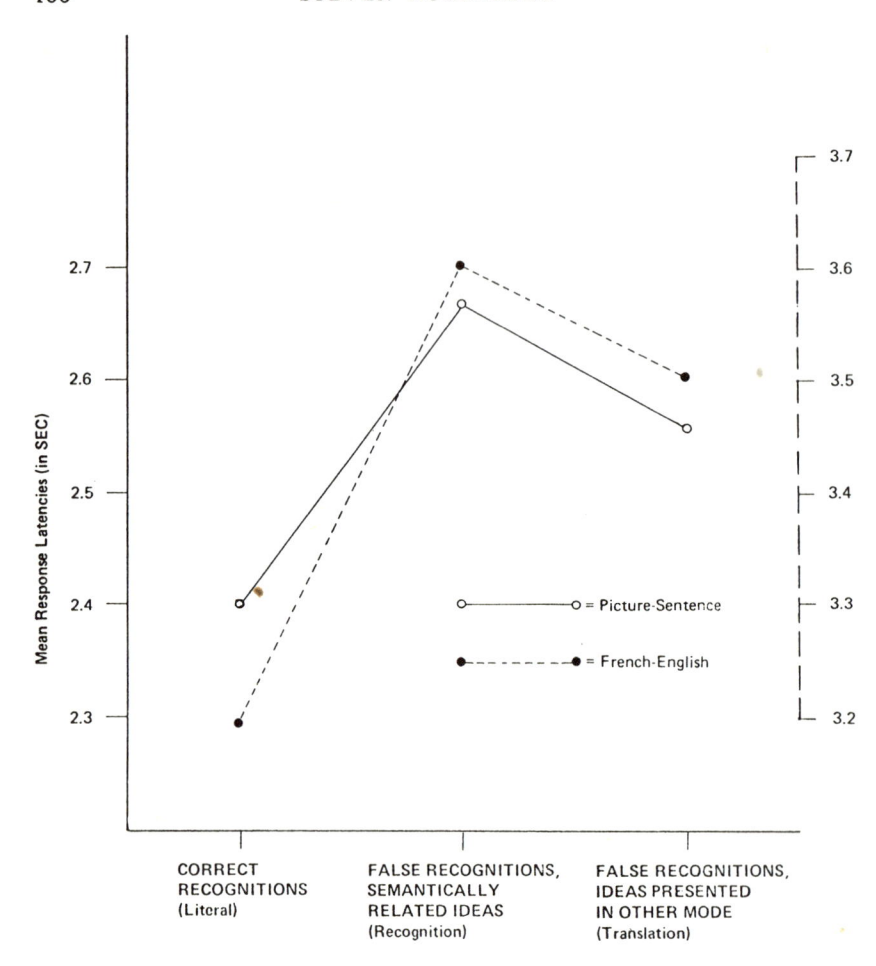

Fig. 3.

pattern of results is the same regardless of the form the material is presented in. At the least, the semantic systems for handling the different forms of the material are very similar, if not identical.

Subjects certainly are not able to reject Translations out of hand. This supports the idea that they integrated information across modalities, using a single semantic system to represent it. However, these results don't suggest a simple system of semantic representation. For instance, not all translated items are falsely accepted. There are differences between the acceptance rates for Literal and Translation items. Thus subjects are able to correctly discern the modality, in some cases, of the original item. On the other hand, the acceptance rates for what should be semantically consistent and acceptable items (Recognition items) are very low. This implies that although subjects integrate related semantic information across modalities, the structure of this semantic representation is complex.

To explain these findings, let us consider the form an information processing model capable of reproducing these results would take.

3. THE MODEL

Our model will need three parts. First we need a system to represent information. Secondly, we will need acquisition procedures for taking in information and adding it to this representation. Thirdly, we require recognition procedures for matching new items against semantic memory. These procedures should integrate related information across modalities and combine it into a single semantic representation, regardless of the modality it occurs in.

3.1 Representation

A modified form of the conceptual case grammar developed by Shank (1972) was used to represent information. The model, however, is not particularly sensitive to the grammar chosen, except that the following semantic distinction must be preserved.

3.2. Predication vs. Attribution

Consider sentences 1 and 2 (and their associated representations). They are examples of predication. Each sentence involves a different type of semantic relation. Sentence 1 contains information that would be added to a 'concept' in a system such as Rumelhart, Lindsay and Norman's (1972). Sentence 2 presents information which would go to an 'event'.

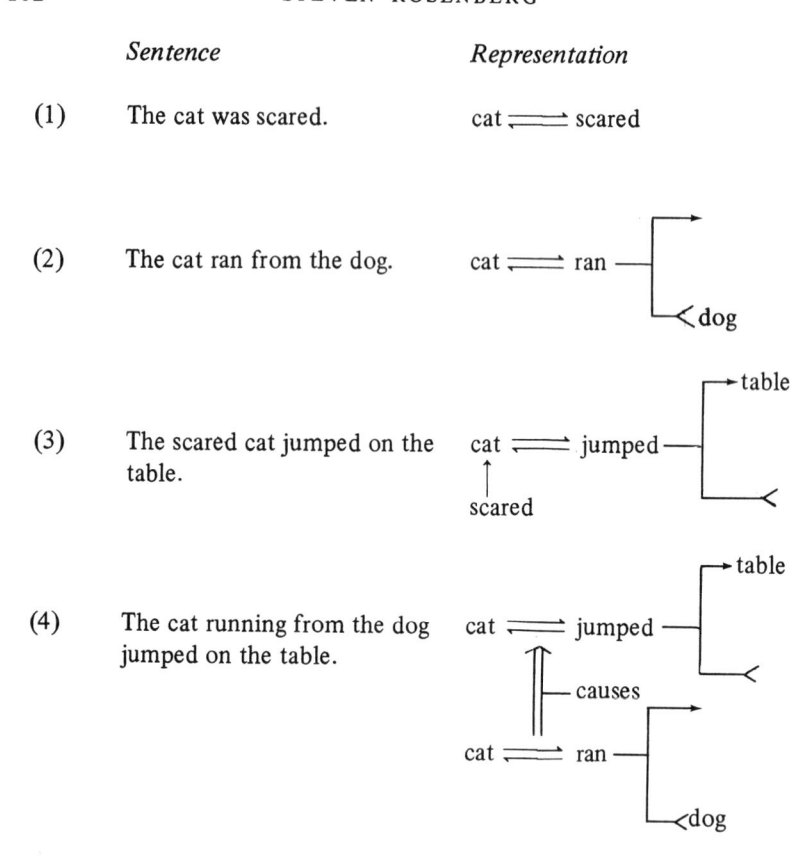

	Sentence	*Representation*
(1)	The cat was scared.	cat ⇌ scared
(2)	The cat ran from the dog.	cat ⇌ ran ─┬─→ └─< dog
(3)	The scared cat jumped on the table.	cat ⇌ jumped ─┬─→ table ↑ └─< scared
(4)	The cat running from the dog jumped on the table.	cat ⇌ jumped ─┬─→ table ↑ └─< ├─ causes cat ⇌ ran ─┬─→ └─< dog

Our model does not differentiate 'concepts' and 'events', but does make a comparable structural distinction when the ideas contained in these sentences are combined with other ideas, to get sentences such as (3) and (4). In both cases, these new sentences have main ideas which are 'events'. The combination of information originally presented as a predication with other information can result in a change in its formal status to that of an attribution. In the case of sentence (1), this change from predication to attribution (sentence (3)) results in a semantic change in the structural representation. Occurring in the predication form, the information was part of a 'concept'; in the new sentence it is made part of an 'event'. This shift results in a change in the way the information is represented. Sentences (2) and (4) show that when 'event' information is made part of a larger 'event' statement there is no change in the representation of that information.

Seemingly equivalent information can then sometimes give rise to different semantic representations. Suppose, for instance, a subject acquired certain information through its occurrence in an 'event' sentence. During a recognition test we present this information as a predication. If it is still an 'event' sentence (as in sentences (4) and (2)) he will have a suitable match onto his semantic representation. If it is a 'concept' sentence (as in sentences (3) and (1)) he will not have an appropriate semantic structure. This might cause a person to reject a sentence although it appears to contain the same information as previously presented material. Such sentences (and pictures) are referred to as those that undergo a predication-attribution structural change.

Since each sentence has a translation as a picture, pictures also undergo this semantic change. Sentences generally undergo a corresponding change in surface sturcture (e.g., 'The cat was scared' to 'The scared cat.....') pictures do not. The same pictorial elements indicate a semantic relation regardless of which other elements are in combination with it. Consequently rejections of sentences undergoing predication-attribution changes cannot be atrributed to a syntactic effect if there are equivalent rejections of pictures. As we shall see, this is what occurs.

3.3. Theme

Another important property of the internal representation is its theme. Every sentence or picture has a theme. For sentences this would be the idea expressed in the main clause; for pictures the idea contained in the main clause of the equivalent translation. As sentences and pictures are integrated into a semantic network the theme idea which has occured most frequently among the integrated items will become the theme of the internal structure. This theme is a pragmatic one, and indicates the central idea of the representation as determined by the input sequence of items. Two different sequences can result in identical semantic structures which differ only on the choice of theme. This can, however, have many consequences.

3.4. Modality

While it is true that our model integrates information across modalities, with semantic relations represented in the same fashion regardless of modality, different modalities will sometimes contain different information. Consider the act of recognizing a picture of a cat, or reading the word 'cat'. To recognize an item we must go to an internal lexicon which contains our knowledge of lexical entries. Each entry defines a type. Every type will have associated with it the possible range of values and properties this type can take. For example, we want to recognize both black and grey cats as being examples of the type 'cat', although they differ on the property of color. Consequently each particular occurence of cat creates a token which is specified for a particular set of properties and their values taken from the potential set associated with the lexical 'type'. This 'type' must have all permissible color values associated with the property 'color'. On seeing a picture of a cat, the resulting token will have the property 'color' with only one value (say black) specified. On recognizing the word 'cat', the property 'color' will be unspecified, but some lexigraphic and linguistic properties may in turn be specified. Consequently the information specified on tokens representing objects may differ from modality to modality. While the specific information associated with a token may vary depending on the modality, the encoding system for information, and the representation of relations between tokens, never changes.

As information is integrated into semantic representations, tokens will often occur several times, sometimes in one modality, sometimes in the other. As a result tokens in a semantic representation may contain properties and values from both modalities. In some cases they will not. Whether they do or not will be a function of the particular items from both modalities which are integrated into a representation. In some cases, a person may be able to tell that an item is a translation because through an examination of the properties of corresponding tokens in his representation he finds he does not have information consistent with that modality for that particular item. This will cause rejection of the translation.

3.5. Acquisition Procedures

Besides a system for representing knowledge, our model requires procedures for the acquisition and integration of information. Figure 4 diagrams the procedures the model will go through when it is presented with new sentences or pictures which are to be added to its semantic memory.

The first element in our diagram is a recognition memory. New input goes to a recognition memory which systematically examines its tokens. Recognition memory contains a dictionary of previously seen tokens. Associated with each token is a pointer to a semantic representation in long term memory. On recognizing some token in a new item, our system

SEPARATION AND INTEGRATION 105

Fig. 4.

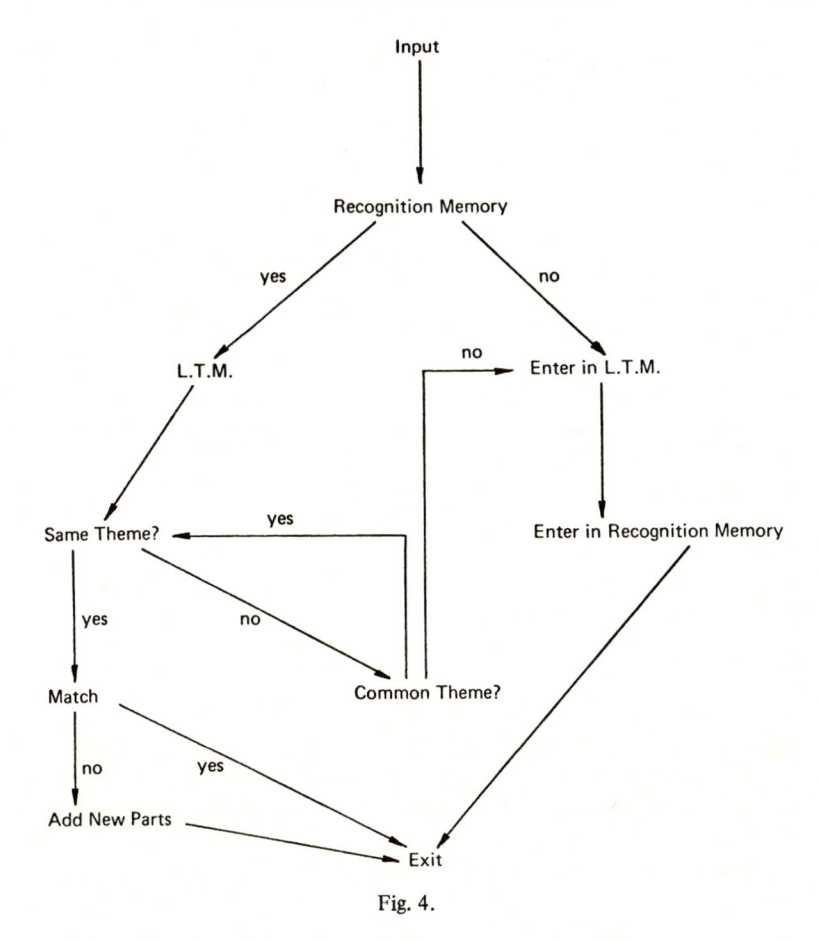

retrieves from long term memory the structure which contains the familiar part.

Successful recognition takes us down the left-hand path in Figure 4, from recognition memory to long term memory, where a semantic structure is retrieved. The model now checks if the new item has the same theme as the retrieved semantic structure. If the answer is yes, the new input is matched against the semantic structure. Wherever the new item contains information not already encoded in the semantic structure this information is added. At this point the system is through with the new item, having successfully integrated it into an existing semantic stucture, and so it exits.

If the two structures do not have the same theme, we go to the node labeled 'common theme'. By a common theme, we mean that the theme of either the new item or the retrieved semantic structure is a non-theme subcomponent of the other. If this is the case, one or the other structure has its theme changed so that both have the same theme; we then follow the same procedure as before.

If this more generous test fails, our system will not seek any further to integrate the new item into the semantic structure. Consequently the theme, representing the central idea among a group of elements, plays an important role in determining when information will be integrated into a semantic structure.

If it is not integrated, the semantic representation of the new item is stored in long term memory as a new semantic structure. A token chosen from the theme idea of the item is entered in recognition memory together with a pointer to its associated representation in long term memory. Similarly, if a new item is not recognized by recognition memory, it will be entered in long term memory, and a new entry is made in recognition memory for it. Each semantic structure has only a single entry in recognition memory.

3.6. Recognition Procedures

Figure 5 shows the processes needed to allow our model to match new items against memory and decide if they have been seen before. Once again, each new item goes to recognition memory first. When this results in a semantic structure being located in long term memory, the new item is matched against this structure. If no semantic structure is found, then of course the new item is rejected. The match of the new item against a semantic representation is such that the representation of each idea in the new item must have an isomorphic match against the corresponding node of the internal representation. The links between ideas must be identical in both structures. The links within ideas of course must also be identical. Tokens must always be of the same type (i.e. two tokens for the type 'cat' for example); however, they need not contain identical information. The information associated with tokens in the new item need only be a proper subset of that associated with the corresponding token in the semantic representation. If this match is successful (This is the *1–1 match* node in Figure 5), then the new item is 'recognized'.

If this match fails, the model tries a weaker match. This is the *1-many*

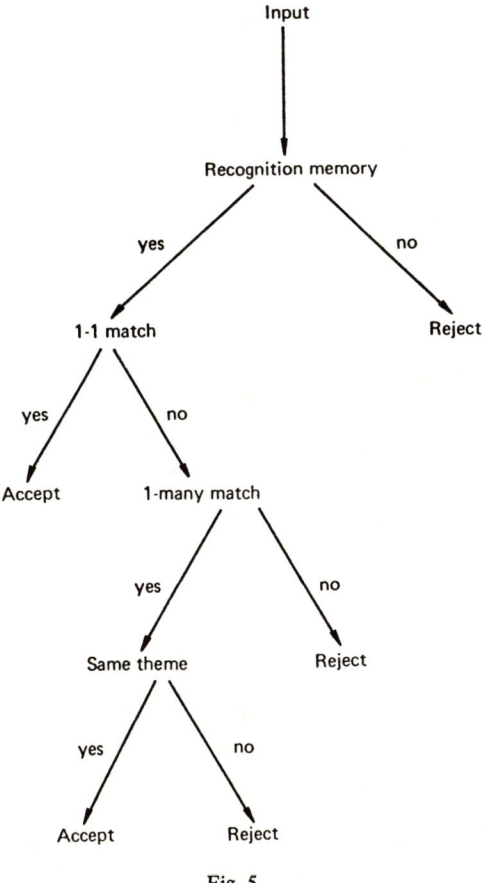

```
                    Input
                      |
                      v
              Recognition memory
               /              \
            yes                no
            /                    \
       1-1 match               Reject
       /      \
     yes      no
     /          \
  Accept    1-many match
              /        \
            yes         no
            /            \
       Same theme      Reject
        /     \
      yes     no
      /        \
   Accept    Reject
```

Fig. 5.

match node. In this case, each idea must be a proper subset of the corresponding idea of the internal representation. Tokens match as before, and the relations between ideas must be identical for the two structures. Since this match is weaker, it is not enough by itself to cause recognition. If it succeeds, the model requires the further step of comparing the themes of the item and the representation. Only if they are identical will it recognize the item. Of course, if this weaker match also fails, the item is not recognized.

The representation, acquisition and recognition procedures contain many

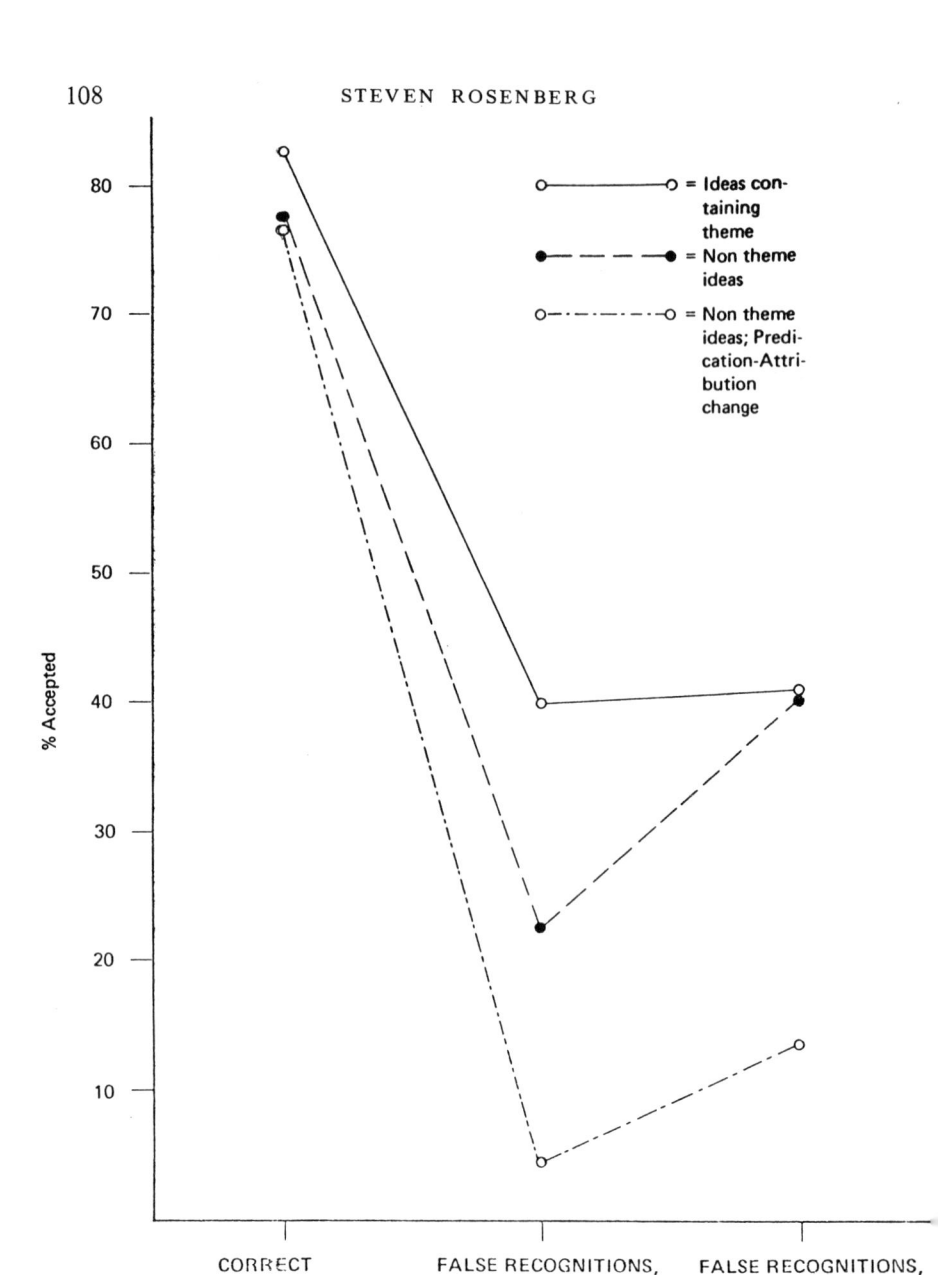

Fig. 6.

features which will cause certain items to be accepted and others to be rejected. There are two aspects which are particularly important: the theme, and the predication-attribution structural change. The theme is a criterion both in adding new information to an existing structure, and in deciding to recognize new items. Sentences and pictures which contain the same theme as a semantic representation should be accepted more frequently than those that do not.

A subset of the non-theme items form a special class. These are the sentences and pictures which can undergo a predication-attribution structural change. When these items occur as Literals, they have the same structure as when they were originally presented. In this case they should not be treated differently from any other non-theme item. When they occur as Recognition items, they will usually undergo a structural change which should result in a mis-match and subsequent rejection. (Since subjects have never seen Recognition items before, their only familiarity with the ideas they contain has been through the occurrence of these ideas combined with others to form different items. Consequently, if these ideas now occur in predication form as Recognitions, they will not match onto their previous occurrences as attributions.)

Similarly, these items may be rejected as Translations when they occur as predications. They can be merged with other ideas only in their attribution structural form. Consequently, tokens in these predications should contain information which allows the modality to be easily distinguished.

Figure 6 shows the data from the Picture-Sentence experiment broken down further into these three new categories (theme, non-theme, and predication-attribution items). Our predictions are in the right direction, with theme containing items being accepted most frequently, followed by non-theme items. Those items undergoing a predication-attribution structural change are accepted the least frequently, particularly when they are Recognitions or Translations.

These particular attributes of the model explain some effects. However, to get quantitative predictions, we need to actually run the entire model and compare its performance to that of subjects.

4. TESTING THE MODEL

We can implement our information processing model on a computer, programmed in LISP to follow the strategies that have been outlined. (See Rosenberg, 1974, for details of this implementation). Let us simulate the

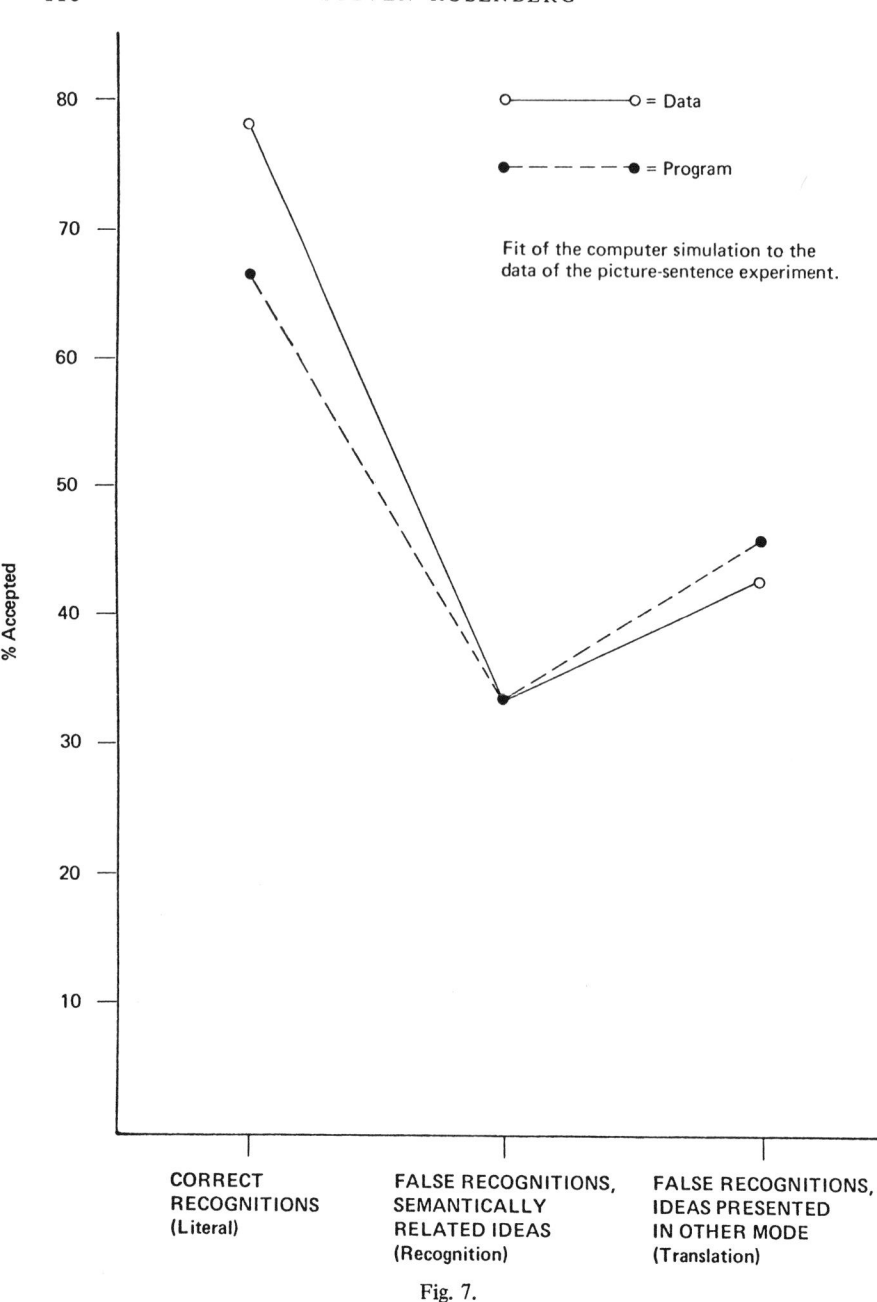

Fig. 7. Fit of the computer simulation to the data of the picture-sentence experiment.

Picture-Sentence experiment by giving the model a subset of the same pictures and sentences to learn and to recognize as subjects received in the various conditions.

Figure 7 shows the match of the simulation to the data when we do so, for the proportions of Literals, Recognitions and Translations accepted. The fit is quite good. Using a single semantic system which merges information across modalities, the model is able to reproduce subjects' correct acceptances of Literals, false acceptances of Translations, and false acceptances of semantically related material (Recognitions). Thus a single semantic system is sufficient to simulate the differences among the levels of Literals and Translations accepted. The lower levels of acceptance for Recognition items are also reproduced, suggesting we have a reasonable model of subjects' semantic domains. By using the power of the computer to calculate outcomes, we have shown how our experimental results can be a function of simple procedures acting on the test material.

The Picture-Sentence study took about twenty minutes for a subject to complete. Over longer time intervals subjects' decision criterion might change. To investigate this possibility another study was performed. The interval between the recognition and acquisition phases was lengthened to an hour, with the whole experiment taking an hour and a half to perform. Only English sentences were used; consequently there are no Translations. Figure 8 shows the results. The data is presented so that Literals and Recognitions can be contrasted for theme, non-theme and predication-attribution change items.

In the Picture-Sentence experiment (Figure 2) there are large differences in the proportions of Literals and Recognitions accepted. After the longer duration of the present study, items which undergo a predication-attribution structural change are still rejected when they occur as Recognitions, and still accepted as Literals. Thus when there is not likely to be a correct matching of semantic structures, subjects' behavior does not change drastically over time. For items for which there is potentially an acceptable match there are changes over time. Initially there was a large (about 45%) difference in the proportions of Literals and Recognitions accepted, with differences between these two groups in the relative proportions of theme and non-theme items accepted. After the longer interval of the present experiment there is now only a small (7%) difference between Literals and Recognitions, with roughly equal amounts of theme and non-theme items being accepted for both groups. There are large differences among the theme and non-theme items, irrespective of whether they are Literals or Recognitions.

Fig. 8.

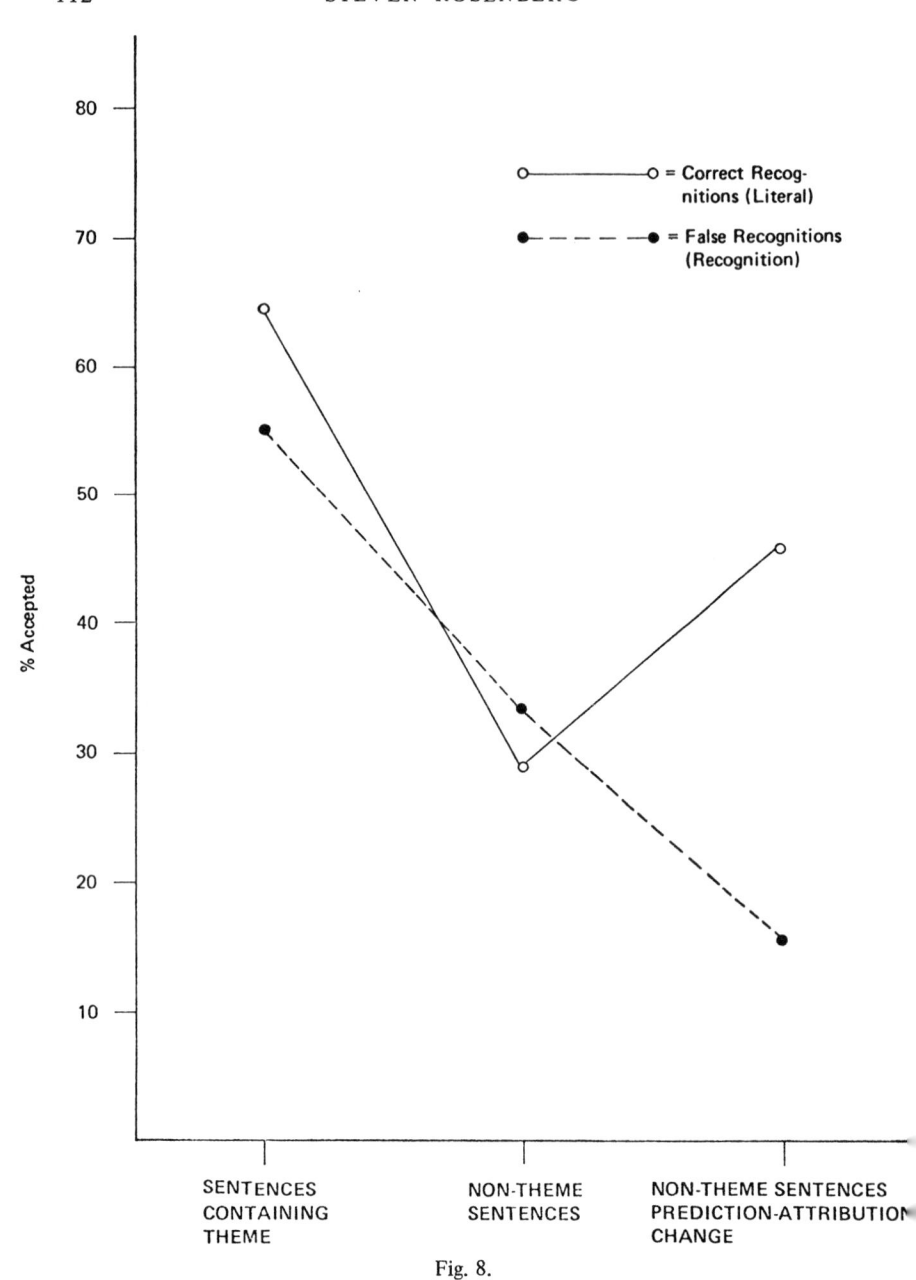

Over time we may have a change not in the semantic representations but in subjects' decision criterion. If our model is robust, it should accommodate changes of this sort. The change in results between the two experiments suggests a simple modification of our recognition procedures. Perhaps the occurrence of the theme becomes so important that this is made the first decision criterion in attempting to recognize an item, rather than the last, as in Figure 5. Figure 9 shows the new arrangement of the recognition processes. Figure 10 shows the fit between our modified model and the data from the new study. The change in the recognition procedure results in a good fit between simulation and data.

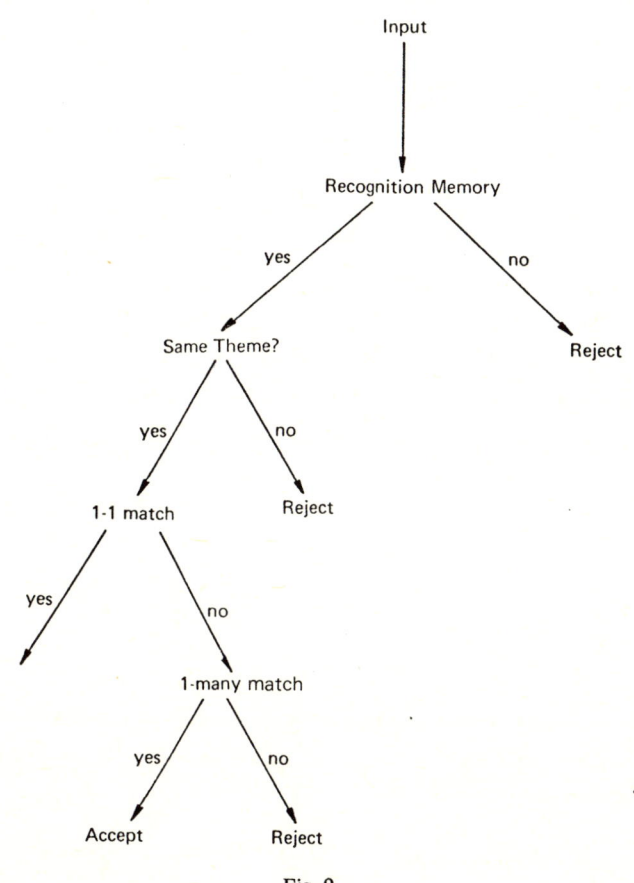

Fig. 9.

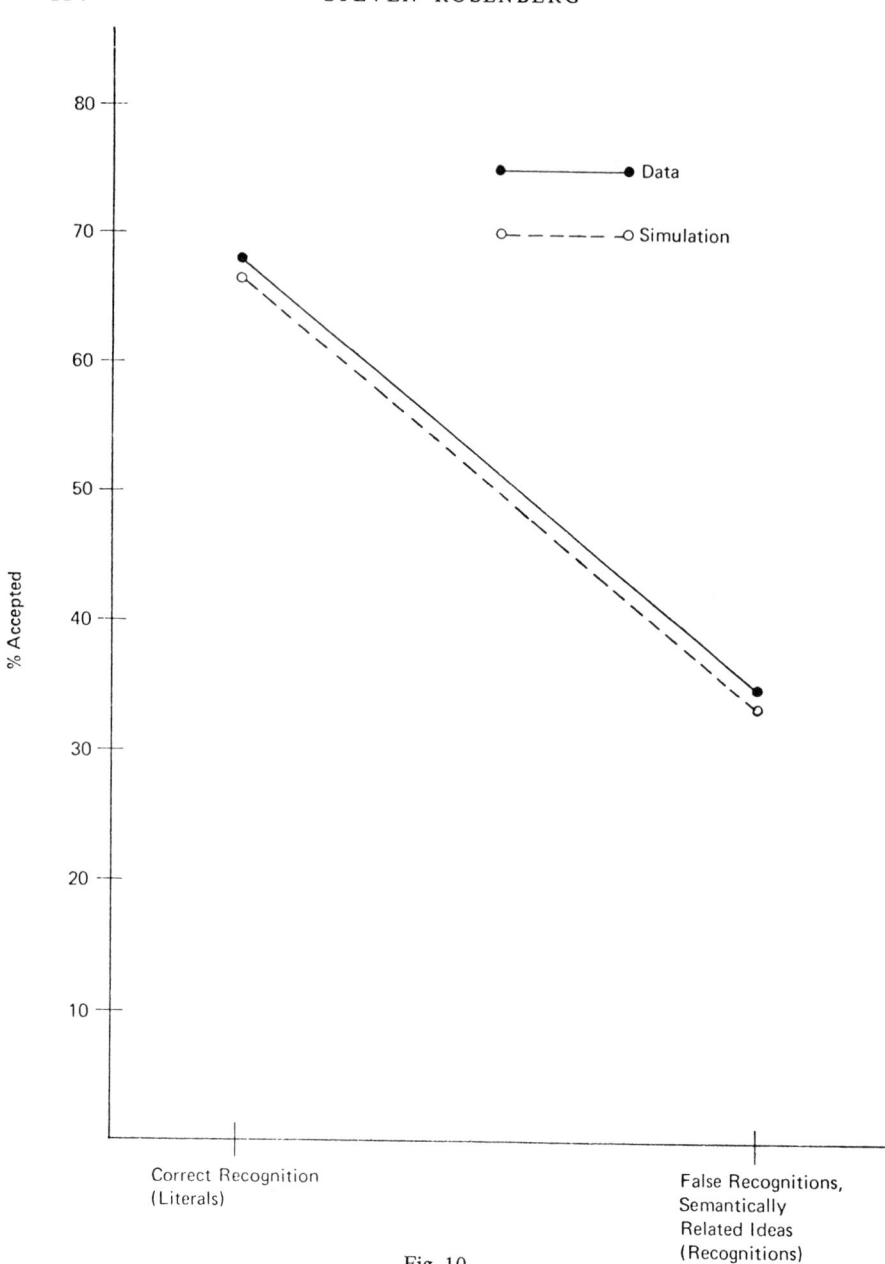

Fig. 10.

The French-English experiment took about 40 minutes from start to finish. Consequently, we can run our revised model on that experiment. Figure 11 shows the match between the model and the data for the particular idea set of sentences available to the model. This group of sentences was used in a control condition where they occurred only in the English language in the acquisition phase. Under these circumstances, the model predicts that all translations should be rejected. It is interesting to observe that even under these ideal conditions about 15% of the translated sentences were still falsely accepted, although subjects had never seen any of the ideas previously in the French language.

Let us now move on to a new semantic domain. We have specified a system for representing information, and acquisition and recognition procedures. A particular semantic domain provides content for the model. However, there is nothing necessarily linguistic, pictorial, or propositional about the system we have built. These are attributes only of the material we give it. Let us investigate how well our model operates in a new domain.

We can create an analogous paradigm to those we used previously by taking letter sequences as our source of material. Thus our two forms might be upper and lower case strings. By using groups of four sequential letters we can create a complex 'idea'. (e.g. *ABCD* or *wxyz*. See Reitman and Bower (1973) for a replication of Bransford and Franks' original experiment which used letter strings rather than sentences.) We can then derive lawful strings, such as *ACD,* or *wx*. *ADC* would be an example of an illegal string.

These simpler materials will not require a complex semantic grammar to represent relations. Simple associative links between adjacent letters will be sufficient. The theme of a letter string can be set to the first letter of the string. Otherwise, we require no changes in our model to handle this new sort of item.

The question we ask with this new material is whether or not subjects store both the physical (i.e. upper or lower case) and name attributes of a string, or only abstract the names of letters, discarding the physical attributes. Of course, we will also be interested in observing if our model can handle the integration of these letter sequences with the same processes as were used for sentences and pictures.

Figure 12 shows the fit between the model and data obtained in preliminary studies. Once again there is an excellent fit for Literals and Recognitions. The model predicts low acceptances of Translations for the sequences of material used in this experiment. Subjects actually accepted about twice as many Translations as predicted. Note however, that they still

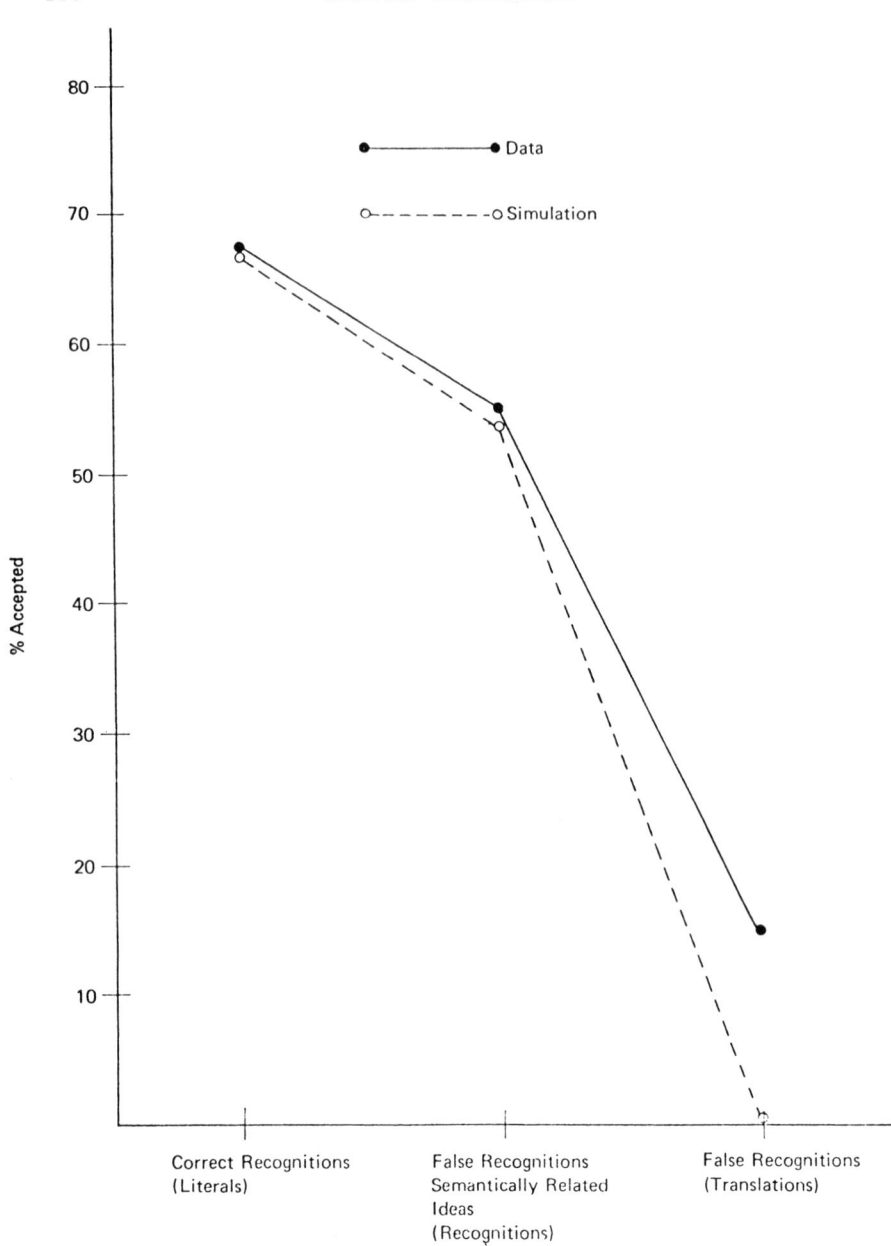

Fig. 11.

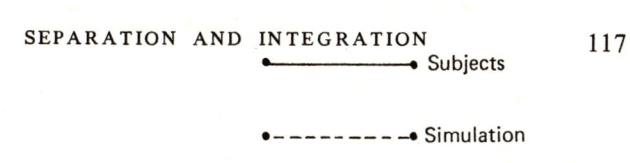

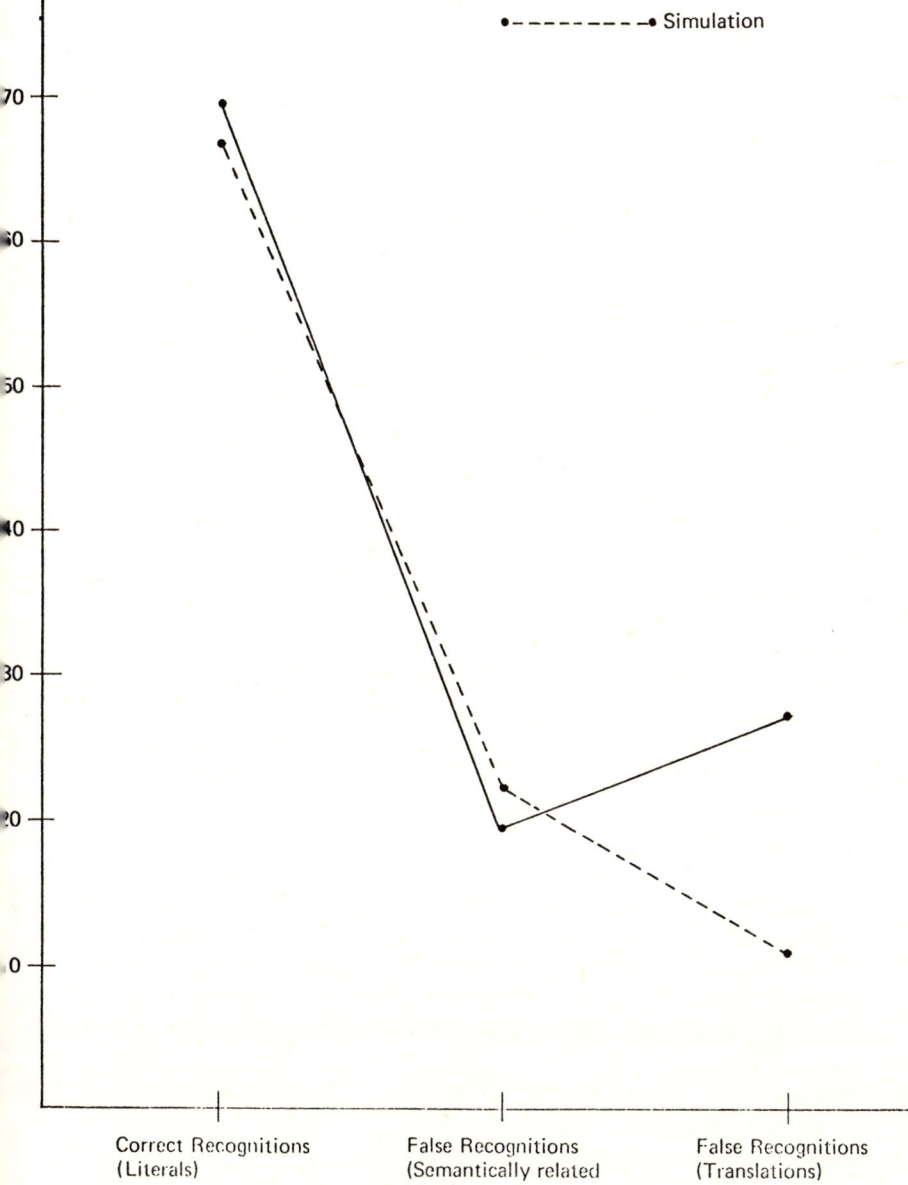

Fig. 12.

do not accept as many as Literals. Thus they do not store only the names of letter sequences, since in many cases they can distinguish the physical form as well.

Our model provides an explanation of how people handle a wide range of material. Different types of material will result in different semantic domains and structures which can yield different quantitative levels of acceptances. The common denominator among the English sentences, French sentences, pictures, and upper and lower case letter strings used, is that in each experiment the same information could occur in more than one form.

We are now in a position to return to the question posed at the beginning of this paper. Do people have different modality-dependent systems for representing information? The answer seems to be that a single semantic system is able to handle results of the sort we have reported here.

A central problem in psychology has been to explain how simple processes can produce complex behaviour. In our case, the power of the computer was used to calculate interactions between procedures. Well defined sets of processes and data representations allowed us to determine what sorts of behaviour would be produced by the model in a given task environment.

These task environments were linked by the fact that in each, the same information could occur in different forms. This produces a natural dichotomy when thinking about a particular study. (For example, linguistic vs. visual encoding; name vs. physical matching.) The approach in this paper has been to avoid extending these categorizations of the experimental situation into presuppositions of an explanatory model. Instead we have asked how can knowledge be represented, and what procedures are required to manipulate it.

We have presented a model of how people integrate related information from different pictures or sentences into semantic representations. This model merged information across modalities, using a single semantic system. This was found to be sufficient to explain the results of several experiments. The present study has hopefully demonstrated how a range of findings may all result from a single set of processes for handling information.

Artificial Intelligence Laboratory
Massachusetts Institute of Technology

NOTE

* Supported by Public Health Services Grant MH-0772 from National Institute of Mental Health.

AARON SLOMAN

INTERACTIONS BETWEEN PHILOSOPHY AND ARTIFICIAL INTELLIGENCE: THE ROLE OF INTUITION AND NON–LOGICAL REASONING IN INTELLIGENCE*

ABSTRACT. This paper echoes, from a philosophical standpoint, the claim of McCarthy and Hayes that Philosophy and Artificial Intelligence have important relations. Philosophical problems about the use of 'intuition' in reasoning are related, via a concept of *analogical representation*, to problems in the simulation of perception, problem-solving and the generation of useful sets of possibilities in considering how to act. The requirements for intelligent decision-making proposed by McCarthy and Hayes are criticised as too narrow, and more general requirements are suggested instead.

1. INTRODUCTION

The aim of this paper is to illustrate the way in which interaction between Philosophy and A.I. may be useful for both disciplines. It starts with a discussion of some philosophical issues which interested me long before I knew anything about A.I., and which I believe are considerably enriched and clarified by relating them to problems in A.I., which, they, in turn, help to clarify. These issues concern non-logical reasoning and the use of non-linguistic representations, especially 'analogical' representations such as maps or models. This discussion is followed by some general speculations about the conceptual and perceptual equipment required by an animal or machine able to cope with our spatio-temporal environment. Finally, there are further vague, general and programmatic remarks about the relations between Philosophy and A.I.

The paper was inspired mainly by discussions with Max Clowes, but also to some extent by the attempts made by McCarthy and Hayes (1969), and Hayes (1969) to relate philosophical issues to problems in the design of intelligent robots. My criticisms of their work should not be taken to imply unawareness of my debts.

Although I was ignorant of the remarkable papers by Minsky while developing these ideas, I now believe that many of his comments on the state of A.I., especially in his 1970 lecture (1970), are intimately connected with the main themes of this paper. I do not yet know enough about computers and programming to understand all his papers listed in my

*John M. Nicholas (ed.), Images, Perception, and Knowledge, 121–138.
First published in: Artificial Intelligence* 2 (1971), 209–225.
Copyright © 1971 by the North-Holland Publishing Company.
Reprinted by permission.

bibliography, so, for all I know, he may already have taken these themes much further than I can.

2. THE LIMITS OF THE CONCEPT OF LOGICAL VALIDITY

Within Philosophy there has long been a conflict between those who, like Immanuel Kant (1958), claim that there are some modes of reasoning, or adding to our knowledge, which use 'intuition', 'insight', 'apprehension of relations between universals', etc., and those who claim that the only valid modes of reaoning are those which use *logically valid* inference patterns. (I shall analyse this concept shortly. The problem of valid inductive reasoning, from particular instances to generalisations, is not relevant to this paper.) Although various attempts have been made to show that non-logical, intuitive, modes of reasoning and proof are important (e.g. I. Mueller (1969) attempts to show that diagrams play an essential role in Euclid's *Elements*), nevertheless, the prevailing view amongst analytical philosophers appears to be that insofar as diagrams, intuitively grasped models, and the like, are used in mathematical, logical or scientific reasoning they are merely of psychological interest, e.g. in explaining how people arrive at the *real* proofs, which must use only purely logical principles. According to this viewpoint, the diagrams in Euclid's *Elements* were stricly irrelevant, and would have been unnecessary had the proofs been properly formulated.

A similar viewpoint seems to prevail in the field of A.I., despite the recent "semantic" approach, which takes non-linguistic models or interpretations into account in attempts to make the search for proofs, or for solutions to problems, more efficient. (For example, Gelernter (1963), Lindsay (1963), Raphael (1968).) The manipulation of non-linguistic structures appears to be tolerated as "heuristics" but not accepted as a variety of valid proof. This prevailing view seems to be implicit in the following quotation from McCarthy and Hayes (1969):

.... we want a computer program that decides what to do by inferring in a formal language that a certain strategy will achieve its assigned goal. This requires formalising concepts of causality, ability, and knowledge.' (p. 463)

Although McCarthy and Hayes do not discuss the question explicitly, their stress on the need for a "formal language" and "formalising concepts", and other features of their essay, suggest that they would not admit the *autonomy* of non-linguistic modes of reasoning. Their concept of a "formal language" seems to include only languages like predicate calculus and programming languages, and not, for instance, the "language" of maps. In his Turing lecture (1970) Minsky inveighed at length against this sort of restriction, but failed to characterise it adequately: it is not, as he suggested, a case of concentrating on form (or syntax) while ignoring content, but a case of concentrating on too narrow a range of types of representations (or "languages"). Formalisation, for instance of syntactic and semantic rules, is indispensable: what is now needed is formalisation of the rules which make non-linguistic, non-logical reasoning possible. I shall support this remark by showing that logically valid inference is a special case of something more general.

What is meant by calling an inference, or step in a proof, from premises $p_1, p_2, \ldots p_n$ to a conclusion c, "valid"? The fact that syntactic tests for validity can be used by machines and by men has led some to forget that what is tested for is not a syntactic relation but a semantic one, which I shall now define.

In general, whether a statement is true or false, i.e., what its truth-value is, depends not merely on its structure, or meaning, but also on facts, on how things are in the world: discovering the truth-value requires the application of semantic interpretation procedures in investigating the world. However, some statements are so related that by *examining* those procedures, instead of *applying* them, we can find that certain combinations of truth-values cannot occur, no matter what the world is like. "London is larger than Liverpool" and "Liverpool is larger than London" are incapable of both being true: they are *contraries* of each other. Similarly some pairs of statements are incapable of both being false: they are *subcontraries*. More generally, when certain combinations of truth-values for statements in some set S are impossible on account of (i) syntactic relations between those sentences and (ii) the semantic rules of the language, then the statements in S are said to stand in a *logical relation*. (A more accurate definition would have to make use of the concept of "logical structure". Although intuitively clear, the precise definition of this concept is very difficult.) Inconsistency, i.e., the impossibility of all statements in the set being true, is one important logical relation. Another is validity of inference, i.e. the case where what is ruled out as impossible is the conclusion, c, being false while all the premisses $p_1, p_2 \ldots p_n$ are true. Thus, logical validity is a special case of the general concept of a logical relation, namely the case where the combination of truth-values $(T, T, \ldots T: F)$ cannot occur.

My main claim is not merely that these are semantic concepts, concerning meaning, reference, denotation (e.g., denotation of truth-values) as well as form (syntax, structure), but that they are special cases of still more general concepts, which I shall now illustrate, with some examples of valid reasoning which are not logical. Many more examples can be found in Wittgenstein (1956).

Consider the familiar use of pairs of circles to represent Boolean relations between classes, as in Figure 1, where (a) represents a state of affairs in which the class A is a subclass of B, (b) represents a state of affairs in which the classes B and C have no common members, and (c) represents A and C as having no common members. If A is the class of male persons in a certain room, B is the class of students in the room and C the class of redheads in the room, then clearly for each of the three figures whether it correctly represents the facts depends on how things are in the world (i.e. in the room). Nevertheless, the "inference" from (a) and (b) to (c) is valid, since no matter how things are in the room, it is impossible for the first two to represent the state of affairs while the last does not: that combination of semantic relations is ruled out, given the "standard" way of interpreting the diagrams. (How is it ruled out?)

Now consider Figures 2a and b, each representing a configuration composed of two horizontal rigid levers, centrally pivoted and joined by an unstretchable string going round a pulley with fixed axle. (A deeper analysis of this example would require a much more elaborate and explicit statement of the semantic rules.) If the arrows represent direction of motion of ends

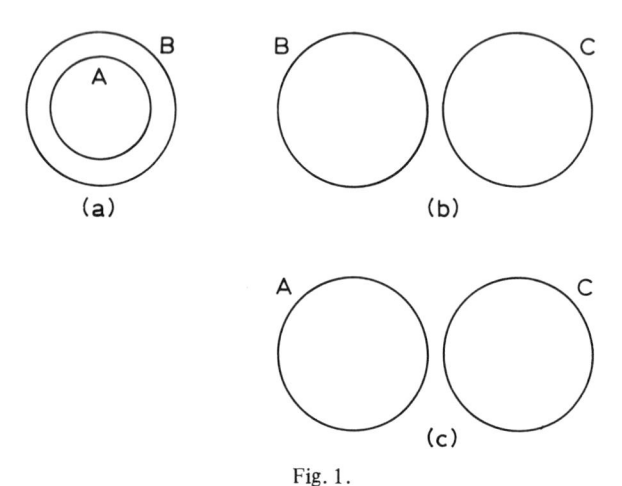

Fig. 1.

(a)

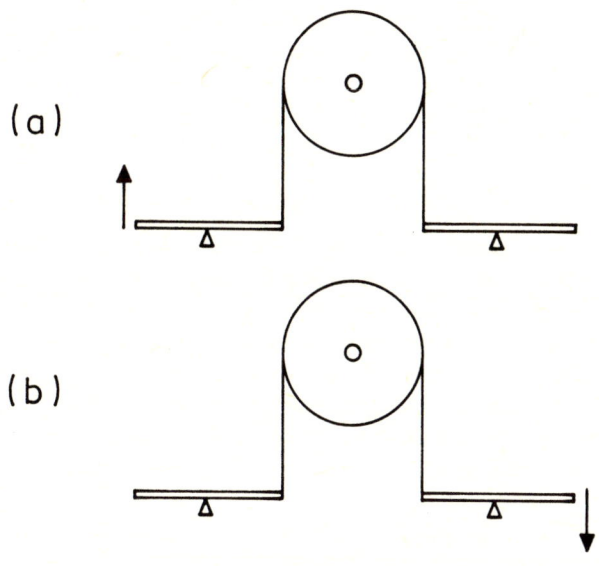

(b)

Fig. 2.

of levers, then it is impossible for any situation to be represented by (a) unless it is also represented by (b), even though whether a particular situation is or is not represented by each of the figures is a matter of fact. Thus the inference from (a) to (b) is valid. Anyone who does not find this immediately obvious may be helped by being shown figures with arrows in intermediate positions, as in Figure 3. This is analogous to the use of a sequence of intermediate steps in a logical proof to help someone see that the conclusion does follow from the premisses: one person may require such

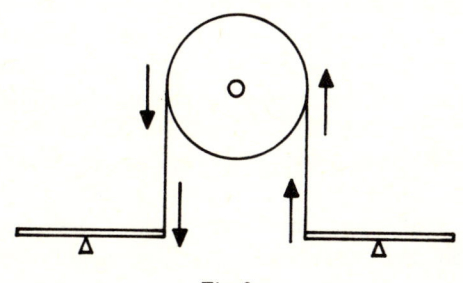

Fig. 3.

intermediate conclusions though another does not. (It would be of some interest to discuss the case of a person who understands each step, but cannot grasp the proof as a whole – but space limitations prevent this. Problem: how do we know where to insert the intermediate arrows?)

3. GENERALIZING THE CONCEPT OF VALID INFERENCE

What these two examples illustrate, is that the concept of a valid inference, or a valid step in a proof, can be generalised in two ways beyond the definition given above. *First* the inference may involve non-verbal representations, instead of only a set of statements. *Second*, validity need not concern only truth-values, but also represented objects, configurations, processes, etc. Thus, the inference from representations $R_1, R_2, \ldots R_n$ to R_c is *valid* in the generalised sense if structural (syntactic) relations between the representations and the structures of the semantic interpretation procedures make it impossible for $R_1, R_2, \ldots R_n$ all to be interpreted as representing anything which is not also represented by R_c. We can also express this by saying that R_c is jointly *entailed* by the other representations.

In this sense (a) and (b) of Figure 1 together entail (c). Similarly (a) in Figure 2 entails (b). Explicitly formulating the interpretation rules relative to which these inferences are valid would be a non-trivial exercise. Once they have been made explicit, the possibility arises of indicating for any valid inference exactly which are the rules in virtue of which the step from "premises" to "conclusion" is valid. When a proof contains such explicit indications it is not merely valid but also *rigorous*. So far, relatively few representational or linguistic systems are sufficiently well understood for us to be able to formulate proofs which are rigorous in this sense. For instance, we can do this for some of the artificial languages invented by logicians, in which various logical symbols are *defined* in terms of their contribution to the validity of certain forms of inference (e.g., the rule of "universal instantiation" is part of the definition of the universal quantifier). But the fact that we do not yet understand the semantics of other languages and representational systems well enough to formulate rigorous proofs does not prevent us from recognising and using valid proofs. Similarly, it need not prevent a robot.

I conjecture that much intelligent human and animal behaviour, including the phenomena noted by Gestalt Psychologists, involves the use of valid inferences in non-linguistic representational systems, for instance in looking at a mechanical configuration, envisaging certain changes and "working out" their consequences. The use and manipulation of rows of dots, or sticks of different lengths, to solve arithmetical problems, instead of the manipulation of equations using numerals and such symbols as "+" and "−" is another example. What philosophers and others have been getting at in talking about our ability to "intuit" or "see" connections between concepts or properties can now be interpreted as an obscure reference to this generalised concept of validity. (My own previous effort (1968–69) was also obscure.) One of the sources of confusion in such discussions is the fact that although we sometimes use and manipulate "external" representations, on paper or blackboards for instance, we also can construct and manipulate diagrams and models "internally", i.e., in our minds. This has led to a certain amount of mystique being associated with the topic. By placing the topic in the context of A.I., we can make progress without being side-tracked into the more fruitless variety of philosophical debate about the ontological status of mental processes, for the ontological status of the internal manipulations within a computer is moderately well understood.

There are, of course, many problems left unsolved by these remarks. For instance, there are problems about the *scope* of particular inference patterns: how far can they be generalised, and how does one discover their limits? (Compare I. Lakatos (1963–64), and S. Toulmin's discussion in (1962) of the use of diagrams in optics.) More importantly, does the ability to generate, recognize and use valid inferences require the use of a "metalanguage" in which the semantic and syntactic relations can be expressed and which can be used to characterise inferences explicitly as valid? Many persons can recognize and use valid inferences even though they have learnt no logic and become incoherent when asked to explain why one thing follows from another: does this imply that we unwittingly use sophisticated metalinguistic apparatus long before we learn any logic? Is *social* interaction required to explain how we can learn the necessary consequences of semantic and syntactic rules? These deep and difficult problems arise as much in connection with the use of language as in connection with the use of non-linguistic representations, so I have no special responsibility for answering them merely because of my defence of non-linguistic systems as having an autonomous status not reducible to the status of heuristic *adjuncts* to linguistic ones.

4. ANALOGICAL VS. FREGEAN MODES OF REPRESENTATION

How should one decide which sort of representational system to use in connection with a given problem? It may be impossible to give a useful general answer to this question, but I shall try to show that for certain sorts of problems "analogical" systems have advantages over general languages like predicate calculus. If this is so, then the hunt for *general* problem-solving strategies and search-reducing heuristics may prove less fruitful than the study of ways in which highly *specific* topic-dependent modes of representation incorporate rich problem-solving powers in their very structures. Contrast Hayes (1969):

... for the robot, generality is all-important, and powerful—problem dependent—heuristics just will not be available.' (p. 536)

Clearly it will depend on the robot: and why should we aim to design only robots whose *general* intelligence surpasses that of humans and other known animals?

In order to make all this more precise we need an analysis of the linguistic/non-linguistic distinction which I have hitherto used without explanation. Detailed investigation shows that there is a whole family of distinctions to be explored. For the moment, I shall merely explain the contrast between "analogical" and "Fregean" modes of representation. Pictures, maps and scale models are largely analogical, while predicate calculus (invented by Frege), programming languages and natural languages are largely, though not entirely Fregean. The contrast concerns the manner in which the parts of a complex representing or denoting configuration, and relations between parts, contribute to the interpretation of the whole configuration, i.e., the manner in which they determine what is represented, expressed, or denoted.

In an *analogical* system properties of and relations between parts of the representing configuration represent properties and relations of parts in a complex represented configuration, so that the structure of the representation gives information about the structure of what is represented. As two-dimensional pictures of three-dimensional scenes illustrate, the correspondence need not be simple. For instance, in Figure 4 distances in the picture represent distances in the scene in a complex context-sensitive way. Similarly, the relation "above" in the picture may represent either "above", or "further", or "nearer" or "further and higher", etc., in a scene, depending on context, such as whether a floor, wall or ceiling is involved. Consequently the interpretation of an analogical representation may involve very complex procedures, including the generation of large numbers of *locally* possible interpretations of parts of the representation and then searching for a *globally* possible combination. For an example see Clowes

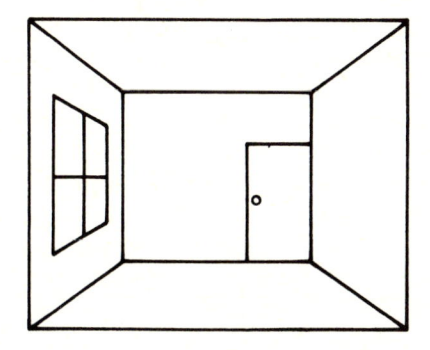

Fig. 4.

(1971). (The use of search-procedures structurally related to the picture is another example of the use of an analogical representation.)

By contrast, in a Fregean system there is basically only *one* type of "expressive" relation between parts of a configuration, namely the relation between "function-signs" and "argument-signs", (Frege's syntactic and semantic theories are expounded in (1960).) For example, the denoting phrase "the brother of the wife of Tom" would be analysed by Frege as containing two function-signs "the brother of ( )" and "the wife of ( )", and two argument-signs "Tom" and "the wife of Tom", as indicated in "the brother of (the wife of (Tom))". Clearly the structure of such a configuration need not correspond to the structure of what it represents or denotes. At most, it corresponds to the structure of the *procedures* by which the object is identified, such as the structure of a route through a complex "data structure". Moreover, the interpretation procedures need not involve the search for a globally consistent interpretation in order to remove local ambiguities, since objects, relations, properties and functions can be unambiguously named by arbitrarily chosen symbols. For instance, the use of the *word* "above" in English need not be subject to the same kind of local ambiguity as the *relation* "above" in Figure 4. Frege showed how predicates, sentential connectives ("not", "and", etc.) and quantifiers could all be used as function signs. Consequently, predicate calculus is purely Fregean, as is much mathematical notation. Natural languages and programming languages, however, are at least partly analogical: for instance, linear order of parts of a programme corresponds, to a large extent, to temporal order of execution. (Devices such as *"go to"* which upset this correspondence are neither Fregean nor analogical. These two categories are by no means exhaustive.)

A Fregean system has the advantage that the structure (syntax) of the expressive medium need not constrain the variety of structures of configurations which can be represented or described, so that very general rules of formation, denotation and inference can apply to Fregean languages concerned with very different subject-matters. Contrast the difficulty (or impossibility) of devising a single two-dimensional analogical system adequate for representing political, mechanical, musical and chemical structures and processes. The *generality* of Fregean systems may account for the extra-ordinary richness of human thought (e.g. it seems that there is no analogical system capable of being used to represent the facts of quantum physics). It may also account for our ability to think and reason about complex states of affairs involving many different kinds of objects and relations at once. The price of this generality is the need to invent complex heuristic procedures for dealing *efficiently* with specific problem-domains. It seems, therefore, that for a frequently encountered problem-domain it may be advantageous to use a more specialised mode of representation richer in problem-solving power. For example, an animal or robot constantly having to negotiate our spatio-temporal environment might be able to do so more efficiently using some kind of analogical representation of spatial structures and processes. A great deal of sensory input is in the form of spatial patterns, and a great deal of output involves spatial movements and changes, at least for the sorts of animals we know about, so the internal decision-making processes involve translation from and into external spatio-temporal configurations. It seems likely, therefore, that the translation will involve less complex procedures, and be more efficient, if the internal representations of actual and envisaged states of affairs, changes, actions, etc., use a medium analogous in form to space-time, rather than a Fregean or other linguistic form.

A great deal more needs to be said about Fregean, analogical and other types of representation or symbol, but I haven't space for an extended survey. Instead I shall now try to describe and illustrate in more detail some ways in which analogical representations may be superior to Fregean or linguistic types.

5. ADVANTAGES OF ANALOGICAL REPRESENTATIONS FOR AN INTELLIGENT ROBOT

An intelligent agent needs to be able to discover the detailed structure of its environment, to envisage various possible changes, especially changes which it can bring about, and to distinguish those sequences of changes which lead to desired or undesired states of affairs. Rumour has it that not all species can do these things in the same contexts: a dog, unlike a chicken, can think of going *round* a barrier to reach food visible on the other side. Similarly, first-generation robots may only have very limited capacities. A minimal requirement for coping with our environment, illustrated by the chicken/dog example, is the ability to consider changes involving relatively smoothly ordered sequences of states, such as going round a fence, turning a knob, moving one end of a stick into the hollow end of another, moving a plank until it bridges a ditch, etc. By contrast, the contexts for intelligent action which appear to have attracted most attention in A.I., such as searching for logical proofs, playing chess, finding a route through a space composed of a network of points and arcs, acting in a world composed of interacting discrete finite automata (compare McCarthy and Hayes, 1969, pp. 470ff), involve search spaces which have no obvious usable order to organisation, so that in order to make problems tractable new organising patterns have to be discovered or invented and new means of representing them created. Of course, these contexts are very important, and are to be found in our environment also (e.g. assembling a mechanism from general-purpose components). But they are also much more difficult, and attempting to tackle them without first understanding how to satisfy the above minimal requirement may be unwise.

For example, here are some problems which we (who? chimps? two-year-old children?) seem able to solve effortlessly when the problem is represented analogically, but which sometimes become much more difficult in a different format (e.g., an arithmetical format, using equations and co-ordinates, etc.). In Figure 2a which way is the right-hand end of the right-hand lever moving? In Figure 5, where A represents the dog, B the food, and the dashed line a fence, find a representation of a route from dog to food which does not go through the fence. (Notice that this requires a grasp of how the latter relation is represented analogically.) In Figure 6, where AB represents a ditch, CD a movable plank, find a way of moving the plank until it lies across the ditch. In Figure 7, representing rail connections

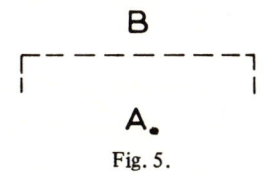

Fig. 5.

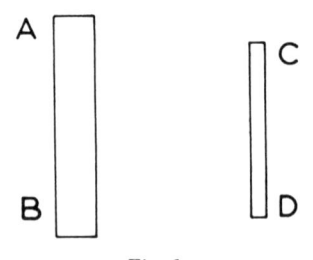

Fig. 6.

between towns, find a route between the two asterisked towns passing through the smallest number of other towns. In Figure 8, where the lines represent rigid rods lying in a plane, loosely jointed at A, B, C, D and E, what will happen to the angle CDE if A and D move together? Our ability to solve such problems "easily" (and many more examples illustrating this could be given) seems to depend on the availability of a battery of "subroutines" which we can bring to bear on parts of spatial configurations, transforming them in specific ways representing changes of certain sorts in *other* configurations.

For instance, while looking at or thinking about some configuration, we can imagine or envisage rotations, stretches and translations of parts, we can imagine any two indicated parts joined up by a straight line, we can imagine X moving in the direction in which Y is from Z, we can imagine various deformations of a line while its ends are fixed, such as bending the line sideways until it passes through some third specified point or until it no longer crosses some "barrier". For example, something like this last procedure could be used to find the route round the fence (see Figure 5), or even a route round a number of barriers — though more than just bending of a straight line may be required if some of the barriers are bent or curved. A

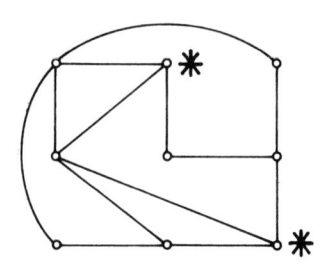

Fig. 7.

Fig. 8.

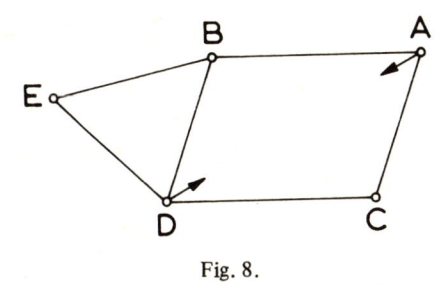

similar routine might be used to find the best route between asterisked points in Figure 7, though more complex procedures are required for more complex route-maps.

Of course we cannot always do the manipulations in our heads: we may have to draw a diagram on paper, or re-arrange parts of a scale model, in order to see the effects. (Try imagining the motion of a worm and pinion without moving your hands.) The difference between performing such manipulations internally and performing them externally is irrelevant to our present concerns. The main point is that the ability to apply such subroutines to parts of analogical configurations enables us to generate, and systematically inspect, ranges of related possibilities, and then, in the generalised sense of "valid" defined previously, to make valid inferences, for instance about the consequences of such possibilities. Thus, in the situation represented by Figure 2a without the arrow, one can find the movement of one lever required for producing desired movement of the other. (Of course, this leaves many unsolved problems, such as how the appropriate manipulations of the representing configuration are selected and how the solution to the problem can be translated into action.)

What these examples seem to illustrate is that when a representation is analogical, small changes in the representation (syntactic changes) are likely to correspond to small changes in what is represented (semantic changes) changes all in a certain *direction* or *dimension* in the representation correspond to similarly related changes in the configuration represented, and constraints in the problem situation (the route cannot go through the fence, the rods cannot stretch or bend, the centres of the pulleys are fixed, etc.) are easily represented by constraints in the types of transformations applied to the representation, so that large numbers of impossible strategies don't have to be explicitly considered, and rejected. Hence "search spaces" may be efficiently organised. By contrast, the sorts of changes which can be made to a Fregean, or other linguistic, description, such as replacing one name with another, conjoining a new description, applying a new function-sign (such as "not-") to the description adding a qualifying phrase, etc., are not so usefully related to changes in the structure of the configuration described. (One can, of course, impose analogical structures on a Fregean system through the use of certain procedures: for instance if names of a class of individuals, are stored in an order corresponding, say, to the order of size of these individuals, then substituting those names in that order in some description, as part of a search, would be similar to the above manipulations of analogical representations. Contrast the non-analogical case where instead of the ordered list, there is a randomly ordered list of *statements* asserting for each pair of individuals which of the two is smaller in size.) For example, "failure of reference" is a commonplace in Fregean and linguistic systems. That is, very many well-formed expressions turn out not to denote anything even though they adequately express procedures for identifying some individual; for example "the largest prime number", "the shape bounded by three straight sides meeting in four corners", etc. (This topic is discussed in my (1971).) It seems that in an analogical system a smaller proportion of well-formed representations can be uninterpretable (inconsistent): pictures of impossible objects are harder to come by than descriptions of impossible objects, so searches are less likely to be wasteful.

A most important economy in analogical systems concerns the representation of identity, or coincidence in complex configurations. Each part of a map is related to many other parts, and this represents a similar plethora of relationships in the region represented by the map. Using a map we can "get at" all the relationships involving a certain place through a single access point, e.g. the dot representing a town whose relations are in question. By contrast, each part of the region would have to be referred to many times, in a large number of statements, if the same variety of information were expressed in linguistic descriptions. (Thus additional semantic rules for identifying different signs as names of the same place are required.) Moreover, a change in the configuration represented, may, in an analogical representation, be indicated simply by moving a dot or other symbol to a new position, whereas very many changes in *linguistic* descriptions of relationships would be required.

Finally, when we use a Fregean or similar language, it seems that our ability to apply names and descriptions to objects in the world has to be mediated by analogical representations. For instance, one can define a word such as "plank" in terms of other words, such as "straight", "parallel", "wooden", etc., but eventually one has to say of some words, to a person who claims not to understand them, "You'll just have to learn how things of that sort *look*". Similarly, any robot using such a language will have to relate it to the world via analogical representations of some sort. So even when deliberation about what to do, reasoning about problems, etc., uses Fregean languages, analogical representations are likely to be lurking in the background, giving content to the cogitations. If so, it may be foolish not to employ whatever relevant problem-solving power is available in the analogical systems.

What I am getting at is that insofar as a robot has to have at least those types of intelligence, common to humans and other mammals, involved in coping with our spatio-temporal environment, it may need to use analogical representations if it is to cope efficiently. Moreover, it should be remembered that although not as general as Fregean representations, spatial analogical representations are useful for a very wide variety of non-spatial systems of relationships, including all those where we find it useful to talk of "trees", "networks", "hierarchies", "spaces" (e.g. search-spaces!), and so on. So the efficient simulation of our sensorimotor abilities may provide a basis for the efficient simulation of a wide variety of more abstract cognitive abilities. (Compare Piaget's speculations about the role of innate motor schemata in cognitive development, reported in Flavell (1963).)

What is now needed is a much more systematic and exhaustive survey of different types of representational systems and manipulative procedures, in order to assess their relative advantages and disadvantages for various sorts of purposes. Some of the ideas in N. Goodman's (1969) may prove useful.

6. SUMMARY OF DISAGREEMENTS WITH McCARTHY AND HAYES

It should be clear by now that although my thinking on these issues has been considerably influenced by McCarthy and Hayes (1969), there are several areas of disagreement, mainly, I think arising out of their neglect of types of representational systems which have not yet been studied by logicians and mathematicians. Where they represent the world as a system of discreet finite automata, I claim that other sorts of representations are more suitable for an environment composed of configurations whose parts and relationships are capable of changing along partially or totally ordered, often continuous, dimensions of different sorts, such as sizes, positions, orientations, speeds, temperatures, colours, etc. Where they analyse the concept of *what can happen or be done* in terms of what is consistent with the interconnections and programs of the automata, I regard this as simply a special case of a more general concept which I call *configurational possibility*, namely the concept of the variety of configurations composed of elements, properties and relationships of the sorts we find in the world. (A fuller discussion would refer to other categories.) Thinking of all the things in one's present environment which might have been bigger, smaller, a different colour or shape, differently located or oriented, moving at different speeds, etc., illustrates the inadequacy of the discrete automaton representation. (Compare Chomsky's proofs of the inadequacy of certain sorts of grammatical theories, in (1957) and elsewhere.) Our ability to notice, and use, such possibilities, apparently shared with other animals, must surely be shared by an intelligent active robot.

Where McCarthy and Hayes require their robot to be capable of "inferring in a formal language that a certain strategy will achieve its goal" (p. 463), I require only that it be capable of recognising a representation of an action or sequence of moves terminating in the goal, and not necessarily a representation in a "formal logical (sic) language" (p. 468). If proof is required that this strategy applied to the assumed existing state of affairs *will* lead to the goal, then a proof within an analogical medium, valid in the generalised sense defined above, will do. Whereas they claim that all this requires "formalising concepts of ability, causality, and knowledge" (p. 463), I claim that it is enough to be able to represent the existing states of affairs, generate (e.g., in an analogical system) representations of possible changes (or sequences of changes), recognize representations of changes which terminate in the goal state, and then attempt to put such changes into effect. There is no need for explicit use of such concepts as "can" or "able" so long as the procedures for generating deformations of representations are geared to what the robot can do. Do dogs and other animals *know* that they cannot do such things as fly over obstacles, push houses out of the way, etc., or do they simply never consider such possibilities in deciding what to do? There is a difference between being able to think or act intelligently and being able explicitly to characterise one's thinking or acting as intelligent. McCarthy and Hayes seem to make the latter a necessary condition for the former, whereas my suggestion is that some of their requirements can be ignored until we are ready to start designing a robot with reflective intelligence.

Of course, a great many problems have been left completely unsolved by these remarks. I have said nothing about how the ability to construct, interpret and modify analogical representations might be programmed. Are new types of computer hardware required if the sorts of subroutines mentioned above for modifying parts of analogical representations are to be readily available? How will the robot *interpret* such routines? How much and what type of hardware and programming would have to be built into a robot from the start in order to give it a chance of learning from experience what its environment is like: e.g. will some knowledge of the form of three-dimensional configurations have to be there from the beginning? Would the ability to cope with some types of *possible changes* in perceived configurations (e.g. motion in smooth curves, rotation of smooth surfaces etc.) have to be programmed from the start in order that others may be learnt?

I cannot answer such questions. What I am trying to do is illustrate the possibility of replacing or supplementing an excessively general and linguistic approach to problems in A.I. with a way of thinking, familiar to some philosophers, involving systematic reflection on facts, about human cognitive abilities, which are readily available to common sense (not to be confused with introspection). By asking, as some philosophers have done "How is it possible for these abilities to exist?", one is already moving in the direction of A.I. The danger is that some people in A.I. pre-occupied with the current technology of the subject and imminently solvable problems may forget or ignore some fruitful new starting points. As for the fear, expressed by Hayes, quoted above, that generality is all-important because powerful problem dependent heuristics will not be available, I hope I have at least given reasons for thinking that they can be made available.

7. PHILOSOPHY AND ARTIFICIAL INTELLIGENCE

Many philosophical problems are concerned with the rationality or *justifiability* of particular conceptual schemes, sets of beliefs, modes of reasoning, types of language. To reformulate these problems in terms of the advantages and disadvantages for an intelligent robot of this or that type of conceptual scheme, type of language, etc., will clarify them and, I hope, stimulate the production of theories precise enough to be tested by using them to design mechanisms whose failure to perform as expected will be a sign of weakness in the theories. Attention paid by philosophers to the problems of designing a robot able to use, or simulate the use of, much of our conceptual apparatus may introduce much greater system and direction into philosophical enquiries (reducing the influence of fashion and historical accidents such as the discovery of paradoxes). I have tried to show, for example, how thinking about the problem of designing a robot able to perceive and take intelligent decisions helps to put logic into a broader context and brings out the importance of storing information in and reasoning with non-linguistic representations: this has important implications also for philosophy of mathematics and philosophy of science. (The sketchiness of some of my arguments is connected with the fact that this paper is part of a much larger enquiry.) Other philosophical problems (the problem of universals, problems about ostensive definition, problems about sense and reference, problems about the relation between mind and body, for example) seem to me to be quite transformed by fairly detailed acquaintance with progress and problems in A.I. This interaction between philosophy and A.I. may also help to remedy some of the deficiencies (such as inept description and explanatory poverty) in contemporary psychology.

School of Social Sciences, University of Sussex

NOTE

* Presented to the 2nd International Joint Conference on Artificial Intelligence, at Imperial College London, September 1971. Since writing it, the author has acquired a keener appreciation of the gap between formulating such ideas and embodying them in a computing system.

D. O. HEBB

CONCERNING IMAGERY

ABSTRACT. An attempt is made to analyze imagery in physiological terms. It is proposed (a) that eye movement has an organizing function, (b) that 1st-order cell assemblies are the basis of vivid specific imagery, and (c) that higher-order assemblies are the basis of less specific imagery and nonrepresentational conceptual process. Eidetic images, hallucinations, and hypnagogic imagery are compared with the memory image and certain peculiarities of the memory image are discussed.

This paper concerns the content and mechanisms of imagery. The topic has received only sporadic attention, partly because of the positivistic temper of modern psychology and partly, one may suppose, because of the difficulties of dealing with thought processes in general. I propose to see what sort of analytical treatment can be made of the image and, equally, of its relation to sensation, perception, and thought. The occasion for such treatment is mainly my interest in thought — one can hardly turn round in this area without bumping into the image — but also the recent work on the place of imagery in paired-associate learning (Paivio, in press) and the convincing demonstrations of eidetic imagery made by Haber and his colleagues (Haber and Haber, 1964; Leask, Haber and Haber) I have also in mind the hallucinatory activity reported in conditions of monotony, perceptual isolation, and loss of sleep (Bexton, Heron, and Scott, 1954; Melvill Jones, cited by Hebb, 1960, p. 741; Malmo and Surwillo, 1960; Morris, Williams, and Lubin, 1960; Mosely, 1953).

1. THE PLACE OF IMAGERY IN OBJECTIVE PSYCHOLOGY

Let me first dispose of what seems to be a misconception, that reporting imagery, or describing it, is necessarily introspective. The point has been made elsewhere (Hebb, 1966) but I repeat it here for those not addicted to introductory textbooks.

An excellent example to begin with is the phantom limb, which is clearly a case of somesthetic imagery. After an arm or leg has been amputated there is, apparently in every case (Simmel, 1956), a hallucinatory awareness of the

John M. Nicholas (ed.), Images, Perception, and Knowledge, 139–153.
First published in: Psychological Review 75 (1968), 466–477.
Copyright © 1968 by the American Psychological Association.
Reprinted by permission.

part that has been cut off. In some 10–15% of the cases the patient also reports pain, the fingers or toes being curled up with cramp. Is this a report of introspection? The argument might be: The pain is in the right hand, but the patient has no right hand; so the pain is really in his mind; so he is describing his mental processes, which is introspection: "looking inward." But the argument is faulty. We are still dealing with a mechanism of response to the environment, though the mechanism (because a part is missing) is now functioning abnormally.

Figure 1 represents a right hand connected schematically with brain and speech organs, before amputation. When the fingers are burnt or cramped the subject (S) says "Ouch" or "My hand hurts." This is a normal mode of response to the environment, involving (a) sensory input, (b) excitation of the central processes of perception and consciousness, and (c) motor output determined by the central activity. It is obvious that in such reactivity – when I burn my fingers and say "Ouch" – no question of looking inward arises. My verbal response is no more dependent on introspection than a dog's yelp when his tail is trod on.

The same conclusion holds after an amputation. No excitation can originate in the missing hand, but the same excitation in principle can arise

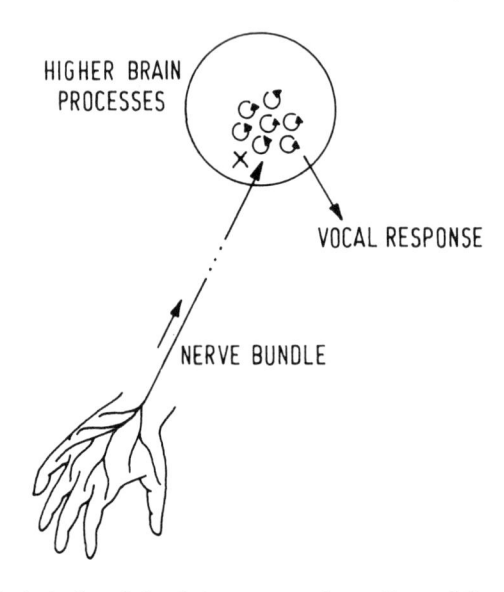

Fig. 1. To illustrate the relation between normal sensation and the phantom limb.

higher in the pathway by spontaneous firing on the neurons at level X in Figure 1. If S now reports pain in his imaginary or imagined hand we are not dealing with any different mechanism, in brain function, than when a normal S reports pain. Report of "sensation" from a phantom limb is not introspective report.

The ordinary memory image can be understood in much the same way. The central processes here may be excited associatively (i.e., the cell assemblies are excited by other assemblies instead of spontaneously firing afferent neurons), but in both cases we are dealing with a short circuiting of a sensory-perceptual-motor pathway. The S on holiday, seeing the ocean for the first time, remarks on the size of the breakers; reminded of the scene later he may say, "I can still see those waves." Though there is now no sensory input, the same central process, more or less, is exciting the same motor response — more or less. (What the differences may be we will consider later.) It is the same outward-looking mechanism that is operative, not introspection.

At least, it is not introspection in the sense of a special inward-looking mechanism of self-knowledge. Anyone may define the term to suit himself, and may use it when reports of private events such as endogenous pain and imagery are in question. My point is that such report does not transcend the rules of objective psychology, in which mental processes are examined by inference and not by direct observation. The primary basis of inference is the relation of overt behavior to present and past stimulation, but there is also a basis of inference about oneself from the appearance of the external world: I may, for example, conclude that I am color-blind if surfaces that others call green and red look alike to me. I also make inferences about the functioning of my visual system when I observe positive and negative after-images though my eyes are shut.

It is important to say also, with regard to a report of imagery, that one is not describing the image but the apparent object. This becomes clear if one observes the apparent locus of what one is describing. One does not perceive one's perceptions, nor describe them; one describes the *object* that is perceived, from which one may draw inferences about the nature of the perceptual process. In the case of imagery, one knows that the apparent object does not exist, and so it is natural to think that it must be the image that one perceives and describes, but this is unwarranted. The mechanism of imagery is an aberrant mechanism of exteroception, not a form of looking inward to observe the operation of the mind. So understood, the description of an imagined object has a legitimate place in the data of objective psychology.

2. WORKING DEFINITIONS

In what follows it will be necessary to distinguish between sensation and perception, without supposing that there is a sharp separation between them. The distinction is based primarily on physiological considerations but the psychological evidence is in agreement. *Sensation* is defined here as the activity of receptors and the resulting activity of the afferent pathway up to and including the cortical sensory area; *perception* as the central (cortico-diencephalic) activity that is directly excited by sensation as defined. For the purposes of this analysis, then, sensation is a linear input to sensory cortex, perception the reentrant or reverberatory activity of cell-assemblies lying in association cortex and related structures.

The term perception itself has two meanings in ordinary usage. Which of the two is intended is usually clear from the context, but when necessary I distinguish *perceiving*, as the process of arriving at a "perception," from a *percept*, the end product, the brain process that is the cognition or awareness of the object perceived. Except with very familiar objects, perceiving is not a one-stage, single-shot affair. It usually involves (a) a sensory event; (b) a motor output, the adjustment of eye, head, or hand to see, hear, or feel better; (c) the resulting feedback; (d) further motor output, further feedback, and so on. As we will see later, this is not a trivial point but must affect our understanding both of percept and of image

Physiologically there is a discontinuity in the mode of operation of the afferent pathway to the sensory cortical area and the structures that lie beyond. The afferent transmission is highly reliable, whereas cortico-cortical transmission, the higher activity that includes perception as defined, occurs only in favorable circumstances. An evoked potential in sensory cortex is obtainable in coma or under anesthesia, but any transmission past this point is not sufficient to break up the synchronous EEG activity. Thus "anesthesia," meaning literally a lack of sensation is a misnomer; we are dealing instead with a failure of transmission at a higher level.

As Teuber(1960) has pointed out, perception cannot be identified with an activity of sensory cortex, so the physiological basis of a distinction between sensation and perception is clear. Sensory systems are organized with fibers in parallel, providing for lateral summation and hence reliability of transmission at each synaptic junction. The divergent course of fibers from sensory cortex onward lacks this feature, and transmission here requires supporting facilitation from the brainstem arousal system, which is absent under anesthesia. The selectivity of response to sensory stimulation even in the normal conscious state strongly indicates that supporting facilitation is also needed from the concurrent cortical activity; except when there is a sharp increase of arousal, due to pain stimulation or certain unfamiliar events, we "notice." perceive, or respond to only those events in the normal environment that are related to what we are thinking about at the moment.

Finally, another relevance of the distinction between sensation and perception from a psychological point of view is the fact that different sensations or sensory patterns can give rise to the same perception, as in the perceptual constancies; and the fact that the same sensory pattern can give rise to quite different perceptions, as in the ambiguous figure (even with fixation of gaze). In this latter case, the only explanation that has been given is that different cell assemblies are excited by the input at different times.

3. THE PATTERN OF ACTIVITY

Both the ordinary memory image and the eidetic image arise from perception. As we will see, this does not mean that the memory image is identical with perception (though eidetic imagery may be), but it does have implications that have not been recognized. The percept of any but the simplest object cannot be regarded as a static pattern of activity isomorphic with the perceived object but must be a sequentially organized or temporal pattern. The same statement, it seems, applies to the memory image.

This has been well established for the image of printed verbal material (Woodworth, 1938, p. 42. cited Binet and Fernald). The S with good visual imagery may be asked to form an image of a familiar word of medium length ("establish" or "material" would be suitable). When he has done so, he is asked to read off the letters backward; or if S is one who reports that when he has memorized verse he can see the words on the page, he may be asked to recall a particular stanza and then to read the last words of each line going from bottom to top. With the printed word before him, spelling the word backward can be done nearly as quickly as forward, but this is not true of the image and the S who tries such a task for the first time is apt to be surprised at what he finds. There is a sequential left-to-right organization of the parts within the apparently unitary presentation, corresponding to the order of presentation in perception as one reads English from left to right and from top to bottom.

Something of the same sort applies with imagery of nonliteral material, though now the order of "seeing" or reporting is less rigid. If the reader will form an image of some familiar object such as a car or a rowboat he will find that its different parts are not clear all at once but successively as if his gaze in looking at an actual car shifted from fender to trunk to windshield to rear door to windshield, and so on. This freedom in seeing any part at will may make one feel that all is simultaneously given: that the figure of speech of an image, a picture "before the mind's eye," in the old phrase, does not misrepresent the actual situation. But Binet (1903) drew attention to a surprising incompleteness in certain cases of imagery, which suggests a different conception. Let us consider the question more closely.

First, consider the actual mechanics of perceiving a complex visual object that is not completely strange but not so familiar that it can be fully perceived at a glance. Figure 2 represents a slightly off-beat squirrel or chipmunk. The eye movements made in perceiving it must vary, but assume that there are four points of fixation, *A, B, C,* and *D*. After fixating these points, perhaps repeatedly, the object is perceived with clarity: *one* percept

Fig. 2. To illustrate the role of eye movement in perception and imagery, (*A, B, C, D,* fixation points.)

is arrived at. But how are the separate visual impressions integrated? We must take account of the fact that these four part-perceptions are all made in central vision, more or less on top of each other, though they are separated in time by eye movements. Each of the four is an excitation of a small group of cell-assemblies, which I will call for the moment Activity A, Activity B, and so on. These activities must take place in the same tissues, more or less intertwined. Activity A is separated from a following Activity B by an eye movement to the right and slightly downward; if Activity D occurs next, it is preceded by an eye movement downward and to the left; and so on. These movements are mechanically necessary in scanning the object, but they may have a further role.

In other words, the motor process may have an organizing function in the percept itself and in imagery. The image is a reinstatement of the perceptual activity, but consider the result if all four of the separate part-perceptions were reinstated at the same time. The effect must be the same as if, in perception, one saw four copies of Figure 2 superimposed to make Points A, B, C, and D coincide, in a mishmash of lines. Instead, Activities A, B, C, and D must be reinstated one at a time, the transition from one to the next mediated by a motor activity corresponding to the appropriate eye movement.

When looking at the actual object each part-perception is accompanied by three motor excitations (assuming these four fixation points) produced by peripheral stimulation. One of them becomes liminal and the result is eye movement followed by another part-perception. If the image is a reinstatement of the perceptual process it should include the eye movements (and in fact usually does); and if we can assume that the motor activity, implicit or overt, plays an active part we have an explanation of the way in which the part-images are integrated sequentially. In short, a part-image does not excite another directly, but excites the motor system, which in turn excites the next part-image. That there is an essential motor component in both perception and imagery was proposed earlier (Hebb, 1949, pp. 34—37) with some informal supporting observations that as far as I can discover are still valid. It is easy to form a clear image of a triangle or a circle when eye movement is made freely (not necessarily following the contours of the imagined figure) harder to do with fixation of gaze while imagining the eye movement, but impossible if one attempts to imagine the figure as being seen with fixation on one point. Though such informal evidence cannot carry great weight, it does agree with the idea that the motor accompaniments of imagery are not adventitious but essential.

4. ABSTRACTION AND HIGHER-ORDER ASSEMBLIES

One of the classical problems of imagery is the generality of the image. Another is its relation to abstract thought. A hypothetical clarification of such questions emerges from a consideration of the relation of secondary and higher-order assemblies to primary ones. There is a classical view going back to Berkeley that an image must be of a specific object or situation and cannot have generalized reference, but Woodworth (1938, p. 43) cites an early paper by Koffka to the contrary, and it seems that the view is more a consequence of theory (regarding the image as reinstated sensation) than of observation. But how can an image be general, or abstract? Again, Binet (1903, p. 124) reports an opposition between thought and imagery. The image, a representative process, is by definition an element in thought: How can we understand such an opposition?

The present status of the theory of cell assemblies is paradoxical, since it has a way of leading to experiments that both support and disprove it. An impressive confirmation from the phenomena of stabilized images (Pritchard, Heron and Hebb, 1960) is matched by quite definite evidence, from the same set of experiments, that the theory is unsatisfactory as its stands (Hebb, 1963). Some of the difficulty for the treatment of perception becomes less with the proposal of Good (1965) that an assembly must consist of a number of subassemblies that enter momentarily now into one assembly, now into another. The assembly itself need no longer be thought of as all-or-nothing in its activity. Fading, for example, may be a function of the density of subassemblies active in a given region, and a strong stimulation may excite all the subassemblies that are available for a given assembly while a weaker stimulation excites fewer of them. A subassembly, conceivably, might be as small as one of Lorente de Nó's closed loops consisting of only two or three neurons.

It is, however, another, aspect of the theory that concerns us now. This is the varying degree of directness of relation between a sensory stimulus and an assembly activity. The old idea that an image must always be of a specific object was the result of thinking (a) that an image is the reinstatement of a sensory-central process, and (b) that the central part of the process corresponds exactly to the sensory stimulation. The epochal work of Hubel and Wiesel (epochal certainly for understanding perception) shows that this may be true for some components of the central process but is not true for others. A "simple cell" in the cortex responds to a specific retinal stimulation, its receptive field permitting little variation; but "complex cells" respond to stimulation in any part of their larger receptive fields, upwards of half a degree of visual angle in extent (Hubel and Wiesel, 1968). A subassembly made up of simple cells, or controlled by them, will thus be representative of a very specific sensory event, but one made up of complex cells will incorporate in itself some degree of generalization or abstraction. Assembly activities accordingly may be more or less specific as perceptual or imaginal events.

The superordinate assembly (Hebb, 1949) takes the process of generalization or abstraction further. The primary or first-order assembly is one that is directly excited by sensory stimulation. The second-order assembly is made up of neurons and subassemblies that are excited, farther on in transmission, by a particular group of primary assemblies; the third-order made up of those excited by second-order assemblies. The theoretical idea is very similar to what Hubel and Wiesel have demonstrated experimentally for simple, complex, and hypercomplex cells; simple cells being those on which a number of retinal cells converge, complex cells those on which simple cells converge, and hypercomplex those on which complex cells converge. From this it may be concluded that the first-order assembly is predominantly composed of or fired by simple cells.

An artificial example to make this specific: Let us say an infant has already developed assemblies for lines of different slope in his visual field. He is now exposed visually to a triangular object fastened to the side of his crib, so he sees it over and over again from a particular angle. Looking at it excites three primary assemblies, corresponding to the three sides. As these are excited together, a secondary assembly gradually develops, whose activity is perception of the object as a whole — but in that orientation only. If now he has a triangular block to play with, and sees it again and again from various angles he will develop several secondary assemblies, for the perception of the triangle it its different orientations. Finally, taking this to its logical conclusion, when these various secondary assemblies are active together or in close sequence, a tertiary assembly is developed, whose activity is perception of the triangle as a triangle, regardless of its orientation.

How realistic is this proposal of complex processes developing further complex processes, in brain function? A heavy demand is made on the brain in the large number of neurons needed for what seems a simple perception. Two comments are in order. As Lashley (1950) observed, the same neuron may enter into different organizations, and many more "ideas" (considered as temporary organizations of neuron groups) are possible than the total number of neurons in the brain; an ideational element may be a phase in a constant flow of changed groupings of neurons. The second point is that there *are* limits on the process of elaboration. A secondary assembly may be the limit of capacity of the rat brain as far as triangles are concerned, for all the complexity of that small brain. The tertiary assembly, it seems, calls for a bigger brain. The rat is doubtfully capable of perceiving a triangle as a whole, even when it has a fixed orientation, and is *not* capable of recognizing a triangle when it is rotated from the position in which he was trained to respond to it. The young chimpanzee, however, or the 2-year-old child, recognizes the rotated figure easily (Gellerman, 1933).

Another example: The baby repeatedly exposed to the sight of mother's hand in an number of positions would develop subassembly and assembly activities corresponding to perceptions of parts of the hand, and then the whole hand, as seen in these varied orientations. As the hand is seen in motion, these assemblies would be made active in close sequence. Their combined effects, at a higher level in transmission would be the basis for forming a higher-order assembly whose activity would be the perception of a hand irrespective of posture.

Although with present knowledge such proposals must be made in very general terms, they are not unreasonably complex in the light of the anatomical and physiological evidence that is available, and they do offer an approach to the otherwise mysterious abstractions and generalizations of thought. Human thought consists of abstraction piled on abstraction, of generalization themselves based on generalizations, and if we are to accept the notion that thought is an activity of the brain we must explore the speculative possibilities of how this may occur.

The actual perception of an object, following this line of thought, involves both primary and higher-order assemblies. The object is perceived both as a specific thing in a specific place with specific properties, and as generalized and abstracted from – but not all of this simultaneously. In imagery, only part of this activity may be reinstated. First-order assemblies, directly excited by sensation, must be an essential feature of perception, but need not be active when the excitation comes from other cortical processes. A memory image, that is, may consist only of second- and higher-order assemblies, without the first-order ones that would give it the completeness and vividness of perception.

5. EIDETIC IMAGERY

We are now in a position to consider a hypothesis of the nature of eidetic imagery. The eidetic image has been regarded with skepticism, I think, because as described it seems to have imcompatible characteristics of both afterimage and memory image. Its occurrence only after stimulation, its transience, and its vividness in detail make it seem like an afterimage; but its apparent independence of eye movement, its failure to move as the eyes move and the possibility of "looking at" its different parts, to see them with equal clarity, means that it cannot be an afterimage. To the skeptic it sounds like an image that has got stuck to the viewing surface, which is unlikely to say the least. Part of the difficulty of understanding disappears if we first assume, with Allport (1928), that the eidetic image is in the same class with the memory image, and if we then recognize that eye movement has a positive integrating role in memory images. Now the scanning of the viewing surface becomes intelligible: As the eidetiker changes fixation from one point to another the motor activity helps to reinstate the corresponding part-percept.

It remains then to account for the detailed vividness of the eidetic image and my hypothesis proposes in short that the eidetic image includes the activity of first-order cell assemblies that are characteristic of perception but absent from the memory image. The idea finds support in the observation of Leask, Haber, and Haber (see their 1969) that the eidetic image may be strictly monocular when formed with one eye open, disappearing when that eye is closed and the other opened. It was proposed above that the first-order assembly is composed of (or controlled by) first-order cells most of which have monocular function (Hubel & Wiesel, 1968).

Since the eidetic image occurs only for a brief period following stimulation, one thinks first of the hypothesis as meaning that there is some after-discharge in the first-order assemblies. But this would imply the continued activity of all of them at the same time, whereas — as we have seen — the activity must be sequential and motor-linked. The hypothesis instead must be that the eidetiker has first-order visual assemblies which for some reason remain more excitable, for a brief period following stimulation, than those of other Ss. It is possible, as Siipola and Hayden (1965) and Freides and Hayden (1966) have suggested, that the differences may be due to some slight brain damage. Other perceptual anomalies suggest, in turn, that one effect of brain damage may be to impair the action of inhibitory neurons in the cortex: neurons whose function is to "turn off" one perceptual (or imaginal) process when it is replaced by another (Hebb, 1960, p. 743). This inhibitory function, however — assuming it exists — is not entirely absent in the Ss of Leask *et al.*, in view of the ease with which the eidetic imagery could be prevented or disrupted.

The disruptive effect of new stimulation (as S looks away from the viewing surface) is intelligible if the subassembly components of the first-order assemblies are now excited in new patterns. Leask et al., also report that "thinking of something else" interferes with the formation of the image, and that the same effect results from verbalization in the attempt to memorize the picture's content. The theoretical implication is that higher-order assembly activity tends to interfere with some lower-order activity if not by direct inhibition then possibly because the higher-order assembly utilizes some of the components of the lower-order one and thus breaks up its organization.

The central fact in this area is Haber's brilliant experimental analysis of eidetic imagery. The present speculation does nothing to extend his results, though it may help to reduce skepticism (for example, in showing how S's eye movement may actively help in retrieving detail). Instead, his work serves here to provide solid experimental data whose import extends to a wider field in which trustworthy data are sparse indeed.

6. HALLUCINATION, HYPNAGOGIC IMAGE AND MEMORY IMAGE

I wish therefore to conclude by taking account of hallucination and hypnagogic imagery, and relating them and the memory image to thought (or, properly, to other forms of thought).

The term hallucination is used here to include any spontaneous imagery that might be taken for a perception, even if S knows that he is not perceiving. A phantom leg, for example, is so convincing that the patient may not realize that the leg has been amputated; at this point it meets the criterion of hallucination in the narrower sense, since the patient is decieved, but it continues to have the same convincing character after the patient is informed of his loss and can see the stump of the limb. The nature of the process has not changed; if it was hallucination before, it is hallucination now. Similarly, the imagery of some Ss in perceptual isolation (Bexton, Heron, and Scott, 1954) was such that they would have thought they were looking at moving pictures if they had not known they were wearing the occluding goggles. This must be hallucination also.

It was proposed above that the memory image lacks vivid detail because it is aroused centrally instead of sensorily. Hallucinations have a central origin also, but their vividness is not inconsistent with the above conclusion. If the cause of hallucinations is spontaneous firing by cortical neurons, the spontaneous firing may occur in first-order as well as in higher-order assemblies. In its vividness, and in its implication of activity by first-order assemblies, the hallucination is like the eidetic image though it is at the opposite pole in its relation to sensory stimulation, since it seems to depend on a *failure* of sensation. In normal waking hours there is a constant modulating influence of sensory input upon cortical activities, helping both to excite cortical neurons and to determine the organization of their firing. When this influence is defective for any reason — pathological processes, or habituation resulting from monotony or "sensory deprivation" — there is still cortical activity. Neurons fire spontaneously if not excited from without. The activity may be unorganized, and in the isolation experiments Ss in fact were in a lethargic state much of the time, unable even to daydream effectively. But when by chance the spontaneous cortical firing falls into a "meaningful" pattern — when the active neurons include enough of those constituting a cell assembly to make the assembly active and so excite other assemblies in an organized pattern — S may find himself with bizarre thoughts or, if first-order assemblies are among those activated, with vivid detailed imagery.

The hypnagogic image, like the eidetic image and unlike hallucination, is an aftereffect of stimulation but there may be a gap of hours, instead of seconds, between stimulation and the appearance of the imaginal activity. It is characteristic of the period before sleep, but on rare occasions may happen at other times. K. S. Lashley once said that after long hours at the microscope watching paramecia he found himself, as he left the laboratory, walking waist-deep through a flowing tide of paramecia (somewhat larger than life-size!) For myself, true hypnagogic imagery is of the same kind though it occurs only before sleep, and is quite different from the ordinary slight distortions of visual imagery at the onset of dreaming and sleep. It depends on prolonged experience of an unaccustomed kind. A day in the woods or a day-long car trip after a sedentary winter sometimes has an extraordinarily vivid aftereffect. As I go to bed and shut my eyes — but not till then, though it may be hours since the conclusion of the special visual stimulation — a path through the bush or a winding highway begins to flow past me and continues to do so till sleep intervenes. The scenes have a convincing realism, except in one respect. Fine detail is missing, I see bushes with leaves on them, for example, but the individual leaf or bush becomes amorphous as soon as I try to see that one clearly, at the same time that its surroundings in peripheral vision remain fully evident. The phenomenon must be very much like the eidetic image, except in its time properties and its lack of fine detail.

The memory image does not share the peculiarities of these other forms of imagery, but it may still be more peculiar than is generally recognized. We have already seen that it lacks detail, due apparently to an inefficiency of associative mechanisms of arousal. We must now observe that the memory image is typically incomplete in gross respect as well. It frequently lacks even major parts of the object or scene that is imagined — though if one looks for them they show up at once and so, unless the question is made explicit, one may have the impression that the whole was present all along. Binet's 14-year-old daughter had the advantage of not being psychologically trained and not realizing how improbable her reports would sound. Asked to consider the laundress, she reported seeing only the lady's head; if she saw anything else it was very imperfect and did not include the laundress's clothing or what she was doing. For a crystalline lens, she saw not the lens but the eye of her pet dog, with little of the head or the rest of the animal; and for a handle-bar, all the front part of her bicycle but missing the seat and the rear wheel (Binet, 1903, p. 126). To think of the memory image as the reinstatements of a single unified perceptual process makes such reports fantastic, but they are not all fantastic when the image is regarded as a serial reconstruction that may terminate before the whole perceptual process has been reinstated.

Incomplete imagery has a special relevance for ideas of the "self" (a mixture of fantasy and realism discussed elsewhere: Hebb, 1960). It is comprehensible of course that one can, with deliberate intention, imagine what one would look like from another point in the room — that is, one can have imagery of oneself as seen from an external point — but a less complete imagery may occur unintended and without recognition. Memory of floating in water commonly includes some visual imagery of water lapping about a face (if one recalls floating face up) or of wet hair about the back of the head (if face down). A long time ago I could introspect with ease and did so freely. Becoming aware that there were theoretical difficulties about introspection, I began to look at the process critically. Eventually I discovered to my astonishment that it included some imagery of a pair of eyes with the upper part of the face (*my* eyes and face) somehow embedded in the back of a head (*my* head) looking forward into the sort of gray cavern Ryle (1949) has talked about. Unfortunately this seemed so ridiculous that I rapidly lost my ability to introspect and now can no longer report on the imagery in detail. But such fantasy in one form or another may be a source of the common conviction that one's mental processes are open to inspection. The imagery is fleeting and unobtrusive and not likely to be reported even to oneself, being so inconsistent with one's ideas of what imagery is and how it works, but it may nonetheless be a significant determinant of thought.

The theoretical analysis earlier in this paper, in terms of lower- and higher-order assemblies, implies a continuum from the very vivid imagery of hallucination throuh the less vivid memory image to the completely abstract conceptual activity that has nothing representational about it. (This includes of course auditory — especially verbal — imagery as well as somesthetic imagery, and it must be wrong to make a dichotomy between visual imagery and thought, or to identify abstract ideas with verbal processes.) The ordinary course of thought involves an interaction of sensory input with the central processes — one looks at the problem situation directly, if it is available, makes sketches, talks to oneself — and the activity of the lower-order assemblies, in imagery, may have the same "semi-sensory" function of modulating the concurrent activity of higher-order assemblies. The relative efficacy of concrete nouns (names of imaginable objects as stimulus-words in paired-associate learning (cf. e.g., Paivio, 1969), together with the fact that pictures of such objects are still better, suggests something of the sort.

Once it has had its effect on higher activity the image may cease; it would be reportable only when it is persistent, tending to interrupt the ongoing thought process, or reinstated later without reinstatement of the whole thought process of which it was part. In this way it is possible to understand how bizarre imagery, of the kind involved in my introspection, might occur without being recognized, or how visual imagery might form an essential part of the cognitive map (Tolman, 1948) of a man driving a car through familiar territory, even for the man who believes that visual imagery plays no part in his planning. The difference between those who have little imagery and those who have much may be not a difference of the mechanism of thinking, but a difference in the retrievability of the image.

McGill University

NOTE
* Preparation of this paper was supported by the Defence Research Board of Canada, Grant No. 9401–11.

KARL H. PRIBRAM

HOLONOMY AND STRUCTURE IN THE ORGANIZATION OF PERCEPTION

1. INTRODUCTION

The face of psychology has undergone a series of changes during a century of growth as a science. Initial concerns with sensory processes, (as, for instance, in the hands of Helmholtz and Mach) and thought (as studied by Külpe, Brentano, and James) gave way to investigations of feelings (e.g., Wundt) and motivations (e.g., Freud). The introspectionism of Tichener was succeeded by the factors of Spearman, Thurston, and Cattell and by the behaviorism of Watson; the Gestalts of Koffka, Köhler, Wertheimer and Metzger were pitted against the learning theories of Pavlov, Hilgard, Hull, Spence, Tolman and Skinner. Each of these faces has left a legacy which can be traced through its descendants and the variety of their modifications, techniques and formal statements of what constitutes psychology, and attests to the vigor of this young science.

During the past quarter century, the ferment has continued. The major influences now are seen to be existential encounter on the one hand and structural analysis based on computers and mathematics on the other. Superficially, it appears as if the earlier apposition of Gestalt to learning theory had gone to extremes: wholism transcendent vs mechanism transistorized. But this would be a superficial reading. A number of transcendentalists are beginning to be seriously concerned with physiological and social mechanisms as explanations of the philosophical teachings of Zen, Tantra and other eastern experiential systems, while, the mechanists have gone cognitive, allowing considerable fluidity and introspective latitude to the models they construct with their computers and mathematics.

The question I want to address, therefore, is whether the time is perhaps ripe for a more comprehensive view of psychological processes — a view that would encompass not only the variety that is psychology, but play a serious role in the scientific Zeitgeist as a whole. Meanwhile, because each current endeavor in psychology, as part of science, is deeply rooted in its technology, the confusion between disciplines continues to be aggravated. Loyalty is often to the discipline or subdiscipline, not to the content of psychology. Thus several groups, though pursuing the same problems, fail to communicate because of the technical jargon developed in each group, even to the use of identical words to convey different referents.

My concern with the problem of disparate theoretical and technical descriptors is a very practical one. I have spent this quarter century performing experiments that purport to relate brain function and behavior to mental processes as these are expressed by verbal (and nonverbal) reports of my fellow humans (often in a clinical situation). In my attempts to communicate the specific fruits of the research results, I have related the function of the frontal cortex of primates to conditional operants; to decisional processes in ROC space; to attention as measured by eye movements, GSR, heart rate changes and reaction time in the presence of distractors; to motivation in relation to food deprivation and pharmacological manipulations; to learning as a functional change in performance; to the structure of memory using computer simulation; and to other brain processes by neuroanatomical and electrophysiological investigations. Intuitively, I feel that what I have found out about frontal lobe function (and limbic system function, and temporal lobe function, etc.) is important not only to brain physiology, but to psychology – and this intuition is shared by most psychologists. Yet in trying to understand and communicate what I have discovered, I come up against a myriad of systems and beliefs: operant conditioners, decision theorists, attention theorists, motivation theorists, learning theorists, memory theorists and neuroscientists of various disciplinary persuasions (e.g., microelectrode artisans, evoked potential analysts, the CNV specialists or EEG computationists, let alone the neurochemists and neuropharmacologists) rarely relate their findings to one another. What is the connection between learning and memory, between attention and decision, between motivation and the various electrical manifestations of brain function? There is no universally agreed answer. It is as if in the physical sciences we did not know the relationship between the moons and their planets, between the solar system and galaxies, between atomic and molecular structure, between mechanical, gravitational and electromagnetic forces.

In short, if I am to make sense of my data, I must come to grips with the multiple framework within which these data have been gathered – the framework we call scientific psychology. This is the task I want to address. Only an outline, a proposal can be entertained in this paper. The detailed fitting of data, working the outline into a coherent body of scientific knowledge will require a more comprehensive effort over the next decades.

The proposal is contained in the holonomic theory. As the name suggests, the theory is holistic. It therefore addresses the interests of Gestalt, of existential concerns, of social encounter and transcendence. However, it is rooted in the disciplines of information, computer and systems analysis and thus aims toward expression of facts in precise mathematical form. The theory, because of its comprehensiveness, has philosophical implications (see e.g., Pribram, 1965, 1971a,b, 1976) but its corpus concerns the relationship of neural, behavioral and experiential levels of inquiry. At this stage, the theory must of necessity be primarily inductive, relying on a systematization of available data and drawing upon metaphor and analogy from more advanced knowledge concerning other physical, biological and social organizations for initial model construction.

In this paper I want, in the tradition of empiricism, to disucss the holonomic theory as it concerns problems of consciousness, perception, imaging and attention, because, as will be shown in the last section of this paper, in a very real sense this area of problems is central to a scientific understanding of anything at all and especially of psychology. My point of departure is brain organization and function as it relates to observations of the behavior (including verbal reports of experience) of the organism in which the brain is functioning. The departure proceeds from a conflict of views which opposes holistic to analytic processes. The following account hopes to show that such opposition is unwarranted, that in fact both types of process occur in the brain and that their interaction is coordinate with perception.

2. THE BRAIN AND THE COMPUTER

One of the most challenging discoveries about brain organization concerns the precise connection between parts of the brain and between these parts and the topography of bodily surfaces. Localization of connections predicts a localization of function. Grossly, this prediction is often confirmed: for example, eyes and ears and nose project by way of nerve tracts to separate parts of the brain and when these parts are damaged, stimulated or electrically analyzed, a correspondence is obtained between anatomical projection and sensory function. The challenge is posed by the precision of the connections. Assignment of a precise function to a particular anatomical arrangement does not come easily. One investigator, Karl Lashley, has even despaired of ever making such assignment and suggested that the anatomy may represent a vestigial residue of some phylogenetically earlier function organization, much as our veriform appendix represents an earlier functional digestive organ (Lashley, 1960).

The problem arises from the fact that large holes can be made in the anatomical organization of the brain without severely disturbing some functions that would be expected to depend on this precise organization. This does not mean that holes in the brain have no effect: when made in the sensory projection areas, for instance, such holes produce scotomata in the appropriate sensory receptive field. However, very little disturbance of sensory, perceptual, attentional, memory or other psychological process can be ascertained when tests are made within the remaining intact field. The remaining brain-behavior field, the remaining neural organization appears capable of taking over, functioning in lieu of the whole — the system shows equipotentiality as Lashley put it (Lashley, 1960). Currently, we would say that sensory input becomes distributed over the reach of the projection system. The question arises, therefore, how.

An alternative to Lashley's phylogenetic argument is to look at current data processing systems for an appropriate analogy. General purpose computers are wired with very specific connections. Yet, one day, in the early period of computer technology, I experienced the following incident: The then current Stanford machine had been sold to a nearby commercial bank to make way for a new installation. Unfortunately, I had collected a batch of irreplaceable data on patients who had received frontal lobotomies some ten years earlier (Poppen et al., 1965), in a tape format tailored to the existing computer. Learning of the replacement only at the last moment, we rushed to the computer center to process our tapes. Much was completed in the next two days and nights, but a small amount of work still needed to be done when, on the third day, dismantling for shipment was begun. We discussed our problem with the person in charge, hoping to delay things by the crucial three or four hours we needed to finish our task. Much to our surprise he said, 'go ahead and keep processing your tapes, we'll begin the dismantling in such a way as not to disturb you'. We were grateful and expected peripherals and cabinets to be tackled first, only to witness the removal of assemblies of switches and tubes from the innards of the machine. Our data processing meanwhile proceeded merrily without any interruption of the cadences to which we had become accustomed. Though we expected the whole affair to come prematurely to a grinding halt at any moment, this did not happen and we gratefully acknowledged the seeming equipotentiality of the man-made brain that had given us such excellent service.

Could it be, that our biological brains, though 'wired' as precisely as any computer, are organized in a similar way — i.e., to be a general-purpose instrument that, when properly interfaced and given proper bootstrap programs to get the 'machine' going, can then handle more complex higher order programs with seeming equipotentiality? Why not? The underlying principles of the operation of biological and hardware brains may be sufficiently similar to warrant such an explanation. An early book with George Miller and Eugene Galanter explored this possibility (Miller *et al.*, 1960) and more recently I presented the neurophysiological and neurobehavioral evidence in support of this approach, pointing out as well, however, the divergences and differences between biological brains and computers (Pribram, 1971a).

One difference involves the very problem of specificity of connections which initiated the present discussion. Computers currently are primarily serial and therefore analytic processors — one event leads to another. Brains, to a much larger extent, are parallel and therefore holistic processors — many related events occur simultaneously.

In an attempt to simulate biological brains on the computer, scientists have constructed programs utilizing highly interconnected hardware which are called random-net configurations. Though these do approximate one aspect of human perception, the constructive aspect (Neisser, 1967), they nevertheless fail when tested against the general characteristics of the human perceptual system (Minsky and Papert, 1969) and fail equally to correspond to the anatomical specificity of the human system in which sensory projections are topologically discrete.

These limitations of hardware simulations have been discouraging to those who felt that current computers were, at least in principle, models of biological brains, and have provided fuel for those who would like to reject the use of mechanistic analogies to the nervous system.

Another interpretation is possible, however. Perhaps we have gained only a partial insight into brain function by stressing essential similarities to the organization of computers. Perhaps what is needed, in principle, is a look at another type of organization conducive to parallel processing, working in conjunction with that represented by present-day computers.

3. THE BRAIN AND THE HOLOGRAM

There is a set of physical systems that meets these requirements — i.e., they display the essentials of parallel processing. These are optical (lens, prism, diffraction, etc.) systems — often called optical information processing systems to distinguish them from the systems of digital switches comprising the computer mechanisms through which programmable information processing is conducted. In optical systems 'connections' are formed by the paths which light traverses and light bears little physical resemblance to the electrochemical energy that is the currency of both brain and computer. Thus the analogy must at once be seen as more restricted. What is to be taken seriously is the analogy between the *paths* taken by the energy, the interactions among these paths and the resulting organizations of 'information' that are produced. Elsewhere, I have, with Nuwer and Baron, discussed possible (and even on the basis of current evidence some probable) physical correspondences between optical and brain systems with respect to these information processing capabilities (Pribram *et al.*, 1974).

The essence of optical information processing systems is their image construction potential. This capacity is to be compared and contrasted with the programming potential of the computer. Neither programs nor images reside as such in the information processing system — they are configurations made possible by the construction of the system. Both images and programs can be captured and stored as such outside their processing systems. When this is done, there appears to be no superficial resemblance between the image or program and the system in which processing takes place, nor even with any readily recordable event structure that occurs during processing. This is because the topography of images and the statements of programs are re-presentations of the process and as such are subject to transformation. The job of the scientist is to specify the transformations that occur between image and optical information processing system and betwen program and computer. The power of these analogies to brain function comes when the mathematical description of these transformations can be shown by experiment to be identical for information processing by the brain as for processing by optical and computer systems. When in addition, the physical components responsible for the transformations are identified, a model of brain function can be constructed and tested deductively by subsequent experiment.

Images and programs are patently different constructions and a good deal of evidence is accumulating to show that in man the right hemisphere of the brain works predominantly in an image mode while the left hemisphere function is more compatible with program processing (see reviews by Sperry, 1974; Milner, 1974; and Gazzaniga, 1970). There is also a considerable body of evidence that this hemisphere specialization is derived from an earlier mammalian pattern of image construction by the posterior-lateral portions of the brain based on somatotopic and visual input, contrasted with a more sequential organization of the fronto-medial (limbic) systems by olfactory and auditory input (see Pribram, 1960 and 1969 for review). These dichotomies are not exclusive and hold only for overall functions — there are many sequential processes involved in image construction (as for instance scanning by the eye of a pictorial array) and there are parallel processes involved in programming (for example, the conducting of a symphony or even the appreciation of auditory harmonics). Yet the fact that neurobehavioral data readily distinguish image and program processing suggests that both must be taken into account in any comprehensive understanding of psychological function.

By contrast to programs, images can be comprehended in their totality even after brief exposures to the energy configurations they represent. They tend to be wholistic rather than analytic, e.g., they tend to completion in the absence of parts of the input ordinarily responsible for them. Also, they tend to be 'good' or 'bad' on the basis of the structure of the redundancy of their components (Garner, 1962). (Programs, on the other hand have no such internal criteria for goodness. A program is good if it works — i.e., is compatible with the computer — and is better if it works faster. When, as in a musical composition, esthetic criteria can be applied, they pertain to the image producing properties of programs, their compatibility rather than their internal structure.) In short, imaging obeys Gestalt principles (which were first enunciated in the visual arts) as would be expected, while programming takes its kinship from linguistics. Both have gained precision and a new level of understanding by recourse to information measurement and processing concepts.

Over the past fifteen years investigating the details of brain function and of psychological processes, in terms of information processing of the programming type, has become reasonably well accepted. Understanding brain function in terms of information processing as in optical systems, leading to image formation, is a more recent endeavor. Yet a sizable body of evidence has accrued to show how parts of the brain are in fact organized so as to construct images.

4. THE EVIDENCE

Much of the recent evidence concerning image formation in the visual system has been provided by Fergus Campbell and his associates. They have established that the visual system is sensitive to the spatial frequencies in patterns of light, much as the auditory system is sensitive to the temporal frequencies in patterns of sound. This sensitivity has been shown both at the cellular level in animals (Enroth-Cugall and Robson, 1966; Campbell et al., 1969a; Campbell et al., 1969b) and in experiments on human psychophysics (Campbell and Kulikowski, 1966; Campbell and Robson, 1968). One of the most important findings from these studies illustrated that the visual system exhibits a systematic tendency to respond to the harmonics of a square-wave grating. This was demonstrated at threshold (Campbell and Robson, 1968) where contrast sensitivity for a square-wave grating was significantly affected by the contrast threshold of its third harmonic, and similarly Blakemore and Campbell (1969) found that adaptation to a fundamental frequency increased threshold for the third harmonic of that frequency. Campbell reasoned, therefore, that the visual mechanism must, much as does the auditory system, decompose any complex wave form into its components, as is done in a procedure developed by Fourier to specify the characteristics of wave forms. Whether in fact the visual mechanism serves as a Fourier analyzer is being tested in several laboratories at the moment by psychophysical experiments (e.g., Stromeyer and Klein, 1974, 1975; submitted). What is necessary is to determine the bandwidth of various channels sensitive to one or another spatial frequency. Campbell's analysis suggested that bandwidths of approximately an octave were involved — a finding consonant with the suggested Fourier mechanism (Blakemore and Campbell, 1969).

These findings have been confirmed and extended in several laboratories. Maffei and Fiorentini (1973) reported that visual cells functioned in fact as Fourier analyzers. Pollen (1971, 1974) determined that the spatial frequency sensitive cells were those that had hitherto been thought sensitive to single bars presented at a certain orientation and that each of these cells responded to approximately an octave of the spatial frequency spectrum. This result was independently obtained by a group of Soviet investigators in Leningrad (Glezer et al., 1973).

There can thus be little doubt that spatial frequency analysis is one function of the visual mechanism. What has this to do with image construction? As already noted, perceptrons have more or less unsuccessfully attempted to make images by the additive sequential and hierarchical process of putting together a figure from the dominant features that compose it. Thus the outlines of a house can be constructed from lines and

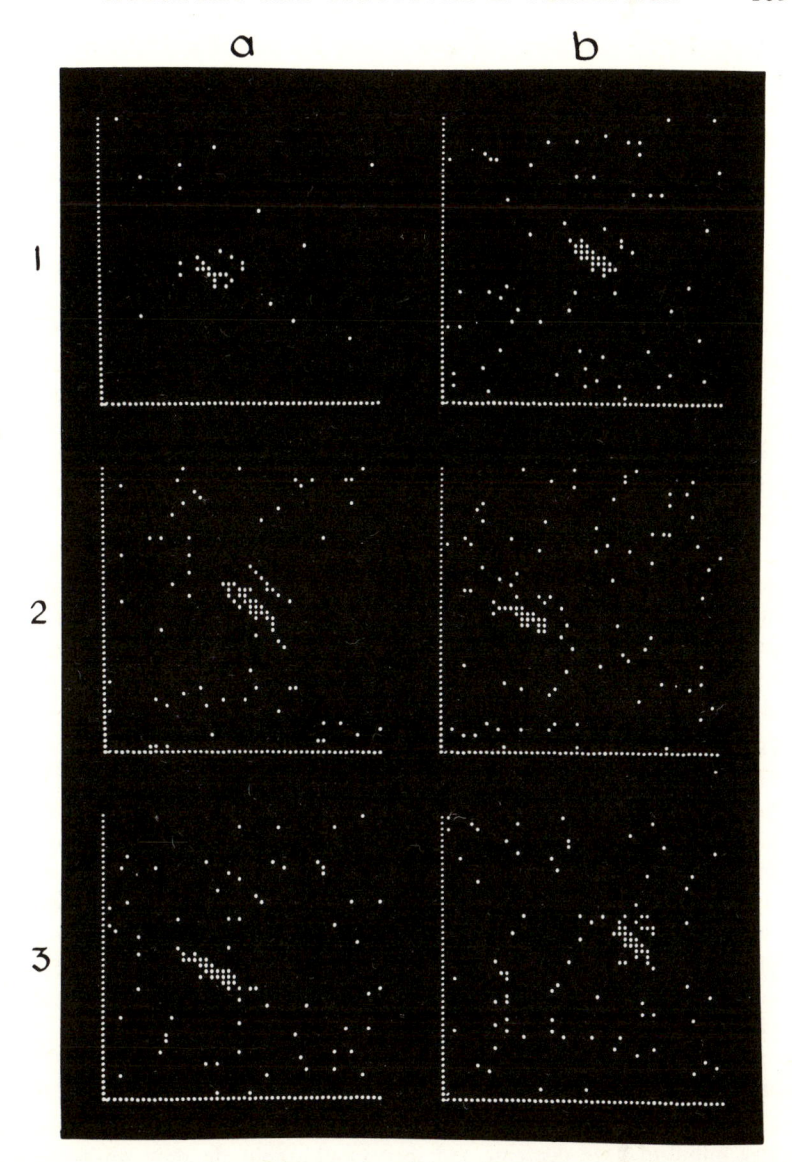

Fig. 1. Visual receptive fields plotted with a moving dot stimulus. (a) right eye; (b) left eye. 1, 2 and 3 are different units. Note the inhibitory flanks next to the main elongated field and that in several of the fields there is a secondary excitatory region.

corners. What is lacking in such a construction is the rich detail, the resolution and fine grain that characterizes our subjective experience of images. This lack is overcome when image construction is based on a spatial frequency mechanism.

Computer simulation highlights the resolving power of the spatial frequency process. Such simulation is performed by composing a figure from square surfaces of different shades of grey (different luminances). It is possible then to manipulate spatial frequencies of different band widths and different dominant frequencies. For example, a crude construction of a face becomes readily reccognizable when the high frequencies that determine the edges of the squares are removed, thus softening the transitions between the contrasting grey areas. Campbell had such a computer analysis and construction performed on a photograph in order to compare the results to those obtained when only lines or only lines and corners were used to make the reconstruction. The results demonstrate conclusively the advantage of the spatial frequency mechanism in providing detail to the image.

How does the brain manage a spatial frequency analysis? Or, for that matter, a temporal frequency analysis? What is the brain process that can perform the transformations necessary to such an analysis whether it be in the Fourier or some similar domain? Neurophysiology has until recently been concerned for the most part with the transmission of signals from one part of the nervous system to another. This transmission is effected by nerve impulses travelling along axons. Transmission is interrupted at axon endings where junctions, synapses, with other neurons occur. Transmission across such junctions is facilitated by the secretion of chemicals at axon endings — neurotransmittors that are stored in vesicles at the presynaptic site.

What has been ignored until lately is the fact that interactions of serious magnitude are occurring among junctional events. Both pre and postsynaptically such interactions block or facilitate conduction of the electrical signal at any particular locus. The interactions can occur because axons branch at their termination and become fibers of small diameter. Postsynaptically, the dendrites leading to the nerve cell body are also fine fibers. When electrical records are made extracellularly from such fine fiber networks of interlacing branches of axons and dendrites, it is found that nerve impulses have decremented into small amplitude slow waves which propagate only short distances if at all. Because of their low amplitude and sluggishness, slow waves are sensitive to local electrochemical fields whether these be generated by neurotransmittors, by the metabolic activities of glia,

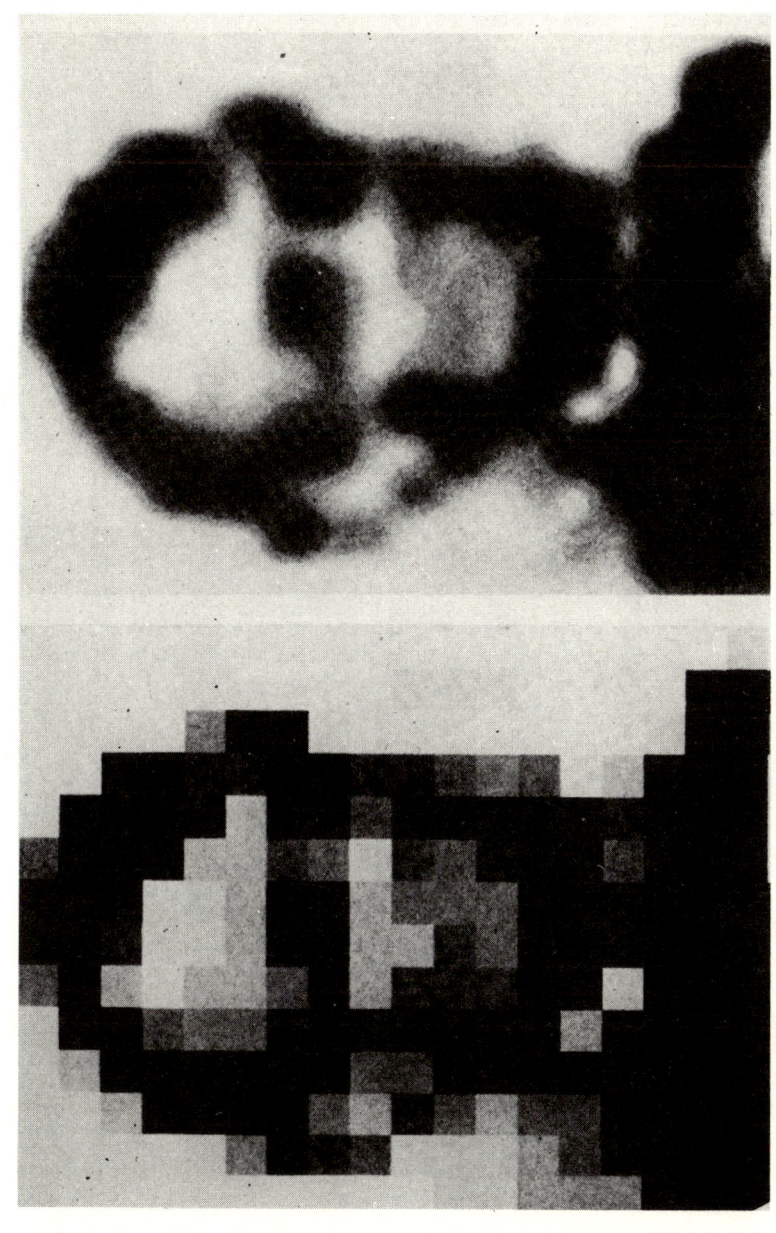

Fig. 2. A computer plot using squares representing different luminances. When high frequency components (edges) are removed, the figure becomes a recognizable portrait.

Fig. 3. A reconstruction of a picture by computer using lines only (right upper), lines and corner (right lower), and various bands of frequency (the four figures on the left). Note the marked improvement in resolution and detail when reconstruction is by the spatial frequency method.

the nutrient supporting cells of the brain, or each other. In short, the pre- and postsynaptic slow potentials can be viewed as constituting an interactive microstructure that has the potential for carrying out the computational work of the brain.

There is agreement among neurophysiologists that a large amount of this computational work occurs at the junctions between neurons — at synapses. By analogy with digital computers and because nerve impulses are discrete events, the workings of the brain have been conceived in digital terms. But by virtue of the interactive nature of the slow potential microstructure, the digital view may be misleading. A view more in keeping with the actual situation would take into account the slow wave nature of the microstructure by the hypothesis that the arrivals of nerve impulses creates a slow wave design — a wave front. This hypothesis would allow the application of wave mechanical mathematics such as Fourier analysis and related techniques (e.g., convolutional integrals, Fresnel and Bessler transforms, etc.) to the study of brain function. The domain of optical information processing would be brought to bear as an important adjunct to the brain's digital programming functions assumed on the basis of integration of information into axonal nerve impulses.

What is the evidence, that in fact, computations by way of a slow potential microstructure do take place? Neuroscientists have come to believe that the most compelling evidence arises from the recent discovery that the computational work of the retina prior to the ganglion cell level is performed exclusively by interactions among slow potentials. No nerve impulses can be recorded from rods or cones, from bipolar or horizontal cells and only rarely from amacrine cells (Werblin and Dowling, 1969). Retinal processes depend on computations performed by a slow potential microstructure. Everything we experience visually is computed by this slow potential microstructure.

The structure of the retina has often been thought to represent a minimodel of the sheetlike portions of the brain such as the cerebral cortex. Microelectrode analysis has supported the view that, at least with regard to the horizontal networks of dendrites (e.g., basal dendrites of the cortex) slow potentials (inhibitory and excitatory postsynaptic potentials) are responsible for the computations reflected in changes of the configurations of receptive fields at progressively more central levels of the visual system (Benevento *et al.*, 1972).

An important consequence of these results of investigation on neural organizations in the visual system is a possible explanation of the mechanism by which input becomes distributed in an essentially parallel processing system. That in fact, such distribution occurs has been shown directly — not only indirectly by sparing of functions after brain resection. Electrophysiological recordings have shown that patterns of electrical potentials evoked by visual stimuli, by responses and their consequences (reinforcements) become separately encoded in a more or less random distribution over the extent of the primate visual cortex (Pribram et al., 1967). This distribution apparently depends on repetition: when nonsense syllables are presented to one retinal locus only once, they are unrecognized when presented at another retinal locus. When, however, such syllables are presented to the same locus several times, they are readily recognized when presented elsewhere (Moyer, 1970).

The data reviewed above taken together with the demonstrations that the visual system is sensitive to spatial frequencies, make it plausible to forward the hypothesis that the interactions among slow potentials, especially in horizontally arranged dendritic networks, are responsible for the distribution of information within the visual system. In optical information processing systems there are loci, planes where the interactions among wave fronts of various spatial frequencies produces interference patterns and a resultant diffusion of information, i.e., information becomes distributed. When these distributed parts of the system (the interference patterns) are captured in a permanent record (as for instance on a photographic film) they are called holograms. By analogy, therefore, the distributed state of information shown to be characteristic in the brain may be called holographic.

5. THE HOLONOMIC THEORY

Holograms provide a powerful mechanism for storing the image constructing properties of optical information processing systems. As already noted, what called attention to the distributed information state is that it makes the brain highly resistant to damage. In addition, the holographic state allows a fantastic memory storage capacity: some hundred million bits of retrievable information have been stored in a cubic centimeter of holographic memory. This is accomplished by separately storing modulations of one or another spatial or temporal frequency. It is somewhat as if there were myriads of FM (frequency modulation) radios compressed into a tiny space. The short wave length of light (as compared to sound) makes such capabilities possible. In the brain, the short wave lengths characterizing the slow potential microstructure can be assumed to serve in a similar fashion.

There are other properties (e.g., associative recall; translational, i.e., positional, and size invariance) of holograms that make the analogy with brain function in perception and memory attractive. These have been presented in another paper (Pribram *et al.*, 1974). Here I want to emphasize that testable hypotheses can be formulated and models of actual brain function can be proposed within the domain of what can loosely be called the holographic properties of optical information processing systems. We have reviewed the evidence for image construction by the brain. What assemblies of neurons (and their processes), if any, function as true Fourier holograms? Which brain structures function more like Fresnel holograms? Which mimic a Fourier process by convolving, integrating neighboring neural events and those at successive stages? These questions are being asked and experiments are being performed to provide answers.

As might be expected, such experiments have already encountered one serious obstacle in drawing too close a parallel between optical information processes and image construction by the brain. This obstacle concerns the size of the receptive fields recorded for cells in the primary visual projection systems. For example, the projection from the macular portion of the retina, the foveal receptive fields, is extremely small — some $3-5°$ of visual angle as a maximum. A hologram of this size will hardly account for the fact that information becomes distributed across the entire visual system, as indicated by the evidence from resections and from electrophysiological recordings.

A search has therefore been made for larger receptive fields that integrate the input from the smaller fields of the primary projection cortex. Such large fields have been found in the cortex that surrounds the primary projection areas. It would be simple if one could assume that here, rather than in the primary projection cortex, the true holographic process takes place.

But this simple assumption runs contrary to other evidence. First, it would not account, by itself, for the distribution of information within the projection cortex. Second, complete resection of this *peri* projection cortex (where the larger receptive fields are found) produces no permanent damage to image construction as far as one can tell from animal experiments (Pribram *et al.*, 1969).

Beyond these visual areas of the brain cortex, however, there is another, lying on the inferior surface of the temporal lobe which, when it is resected, leaves monkeys markedly and permanently impaired in their ability to make visual discriminations (Pribram, 1954, 1960, 1969). This impairment is limited to the visual mode (Pribram and Barry, 1956; Wilson, 1957). Only visual performances demanding a choice are impaired; other visual functions, such as tracking a signal, remain intact (Pribram, Chapter 17, 1971a). The difficulty involves the ability to selectively attend to visual input (Gerbrandt et al., 1970; Rothblat and Pribram, 1972; Gross, 1972).

Much to everyone's surprise, this visual 'association' area (as the area with comparable function is known in man: Milner, 1958) appears to function remarkably well when all known visual input to it is destroyed. As already noted, removal of the perivisual cortex has little permanent effect; destruction of the thalamic input (from the pulvinar) to the inferior temporal cortex has no effect whatsoever (Mishkin, 1972; Ungerleider, personal communication). Even combined lesions of perivisual and thalamic inputs do not permanently disrupt visual discriminations.

These data make plausible the hypothesis that the inferior temporal cortex exerts its effect on vision via an output to the primary visual projection system (Pribram, 1958). Evidence in support of this hypothesis has accrued over the past fifteen years: the configuration and size of visual receptive fields can be altered by electrical stimulation of the inferior temporal cortex (Spinelli and Pribram, 1967); recovery cycles in the visual projection systems are shortened by such stimulation (Spinelli and Pribram, 1966); the pathways from the inferior temporal cortex have been traced (Whitlock and Nauta, 1956; Reitz and Pribram, 1969).

Thus, another, more specific hypothesis can be entertained – viz., the suggestion that the inferior temporal cortex helps to program the functions of the primary visual projection systems. Specifically, such programming, as well as programming by input from sensory receptors, could 'get together' the distributed store of information from the various loci of restricted receptive field size. If the relevant loci were addressed in unison they would, in fact, function like a hologram.

The difference, therefore, between brain function and the function of optical information processing systems is the one set out at the beginning of this paper. Brain is *both* an image construction and a programming device. Optical systems construct only images.

The thesis presented here, therefore, suggests that the holographic-like store of distributed information in the primary visual projection systems is akin to the distributed memory bank of a computer. The computer's memory is organized more or less randomly; the brain's memory has been stored along holographic principles. Both must be addressed by programs which access the appropriate 'bits' of information. The computer does this serially; the brain, to a large extent, simultaneously, by pathways that allow signals to be transmitted in parallel. Such simultaneity in function produces momentary brain states that are akin to the holographic patterns that can be stored on film.

Because of these differences between brain and optical systems, it may be better to talk about brain function as holonomic rather than just holographic or hologrammic. The term holonomic is used in engineering whenever the systems, in an interactive set of such systems, are reasonably linear in their function. Linearity allows the computation of the functions of each system and therefore an estimate of the amount of their interaction — the 'degrees of freedom' that characterize the interactive set. The interactions are known as the holonomic constraints on the system. In the context of the model of brain function in vision suggested here, the neural systems that determine any momentary visual state would have to be shown to be linear; then the amount of interaction among the systems in producing the holographic visual state would appear as the degrees of freedom characterizing that state.

Evidence is available to show that the visual system, despite local non-linearities, acts linearly overall above threshold (e.g., Ratliff, 1965). This is the case in other neural systems, notably the motor system (Granit, 1970). It is thus reasonable to propose that the holonomic model applies to brain functions other than visual. Support for such a proposal comes from work on the auditory (von Bekesy, 1960), somatosensory (von Bekesy, 1959) and even gustatory (von Bekesy, 1967; Pfaffmann, 1960) and olfactory systems (Gesteland *et al.*, 1968).

Briefly summarizing, the holonomic model of brain function proposes that the brain partakes of both computer and optical information processes. The brain is like a computer in that information is processed in steps by an organized and organizing set of rules. It differs from current computers in that each step is more extended in space — brain has considerably more parallel processing capability then today's computers.

This parallel processing aspect of brain function leads to another difference. The rules of parallel processing are more akin to those that apply to optical information processes than they are to those used in current serial computers. Thus the momentary states set up by the programming activity are considerably like those of image constructing devices, i.e., holographic. Thus memory storage is also holographic rather than random as in today's computers. This does not deny, however, that storage of rules also takes place — as it does in machine peripherals (e.g., DEK tapes for minicomputers). What the model requires is that the 'deep structure' of the memory store is holographic.

Since the holographic state is composed by programs and since the distributed store must be got together by the actions of and interactions among programs, the holographic brain state can be analyzed according to the systems that produce it. Thus the holonomic constraints or degrees of freedom that characterize the holographic state can be determined. The holonomic model of brain function is therefore mathematically precise, and its assumptions (such as the overall linearity of component programming systems) and consequences (the distributed nature of the deep structure of the memory store) are, at least in principle, testable.

6. IS PERCEPTION DIRECT OR CONSTRUCTIONAL?

I want now to address some consequences to psychology (and perhaps to philosophy) of the holonomic theory of brain function. The theory, as we have seen, (1) stems from the metaphors of machine and optical information processing systems; (2) has developed by analogy to those systems, spelling out some similarities and some differences; until (3) a testable holonomic model of brain function could be proposed. One way of understanding the model better is to compare it to another and to observe its relative explanatory power.

An apparent alternative to the 'holonomic' model is presented by James Gibson's comprehensive 'ecological' model of perception (1966). Gibson's model proposes that the 'information' perceived is inherent in the physical universe and that the perceiver is sensitive to whatever information remains invariant across transformations produced by changes in the environment, by organism-environment displacements, and by the organism's processing apparatus. The key concept in the ecological theory is 'direct perception' — the environment is directly apprehended by the perceiver.

By contrast, the holonomic theory is constructional. Images are constructed when input from inferior temporal cortex (or its analogue in other perceptual systems — see Pribram, 1974a) activates, organizes the distributed holographic store. Images are produced and are therefore as much a *product of* the 'information residing in' the organism, as they are of 'information' contained in the environment. Philosophically speaking, the holonomic model is Kantian and Piagetian, the ecological model partakes of a naive realism.

Clinical neurological experience wholly supports the holonomic view. Patients are seen who complain of macropsia and micropsia and other bizarre distortions of visual space. For instance, I once had a patient who, after a blow on the head, experienced episodes of vertigo during which the visual world went spinning. His major complaint was that every so often when his perceptions again stabilized, they left him with the world upside down until the next vertigo which might right things once again. He had developed a sense of humor about these experiences, which were becoming less frequent and of shorter duration: his major annoyance he stated to be the fact that girls' skirts stayed up despite the upside-down position!

Further 'clinical' evidence in support of the holonomic model comes from the experimental laboratory. Resections of the primate peri-visual cortex markedly impair size constancy — the transformations across various distances over which environmental information must remain invariant in order to be 'directly' perceived as of the same size (Ungerleider, Gauz and Pribram 1977).

Yet Gibson (1966, 1968) and others who share his views (e.g., Johansson, 1973; and more recently Hebb, this volume), make a good case that, in normal adult humans, perception is direct. A series of ingenious experiments has shown that by appropriate manipulations of 'information', illusions indistinguishable from the 'real' can be created on a screen. The demonstrations are convincing and make it implausible to maintain a solopsistic or purely idealistic position with respect to the physical universe — that nothing but a buzzing blooming confusion characterizes external reality. With respect to the experiments he has devised, Gibson is correct.

Furthermore, if perception is direct, a dilemma for the holonomic theory would be resolved. When an optical hologram produces an image, a human observer is there to see it. Whan a neural hologram constructs an image, who is the observer? Where is the 'little man' who views the 'little man'? Direct perception needs no little men inside the head. Gibson, in fact, (1966) deplores the term image because it calls up the indirectness of the representational process. However, if what we 'directly perceive' is a constructed *image* and not the true organization of the external world — and we mistake this perception as veridical — perception would be both direct and constructional.

The question to be answered therefore is by what mechanism can perception be both direct and constructional? A clue to the resolution of the dilemma comes from the Gibson (and Johansson) experiments themselves. Their displays produce the *illusion* of reality. When we know the entire experiment we can label the percept as an illusion, even though we directly experience it. In a similar fashion, the sound coming from the speakers of a stereophonic system is experienced directly. When we manipulate the dials of the system (changing the phase of the interacting, interfering sound waves) so that all of the sound comes from one of the speakers, we say the speaker is the source of the perception. When we manipulate the dials so that the sound emanates from somewhere (e.g., the fireplace) between the speakers, we say that an illusion has been produced — the sound has been projected to the space between the speakers. Perception continues to be direct, but considerable computation is involved in determining the conditions over which the 'information' contained in the sound remains invariant. We do not naively assume that the fireplace generates the sound. Despite the directness of the perception, it can be superficially misleading as to the actual characteristics of the physical universe.

The issues appear to be these. Gibson abhors the concept 'image'. As already noted, he emphasizes the 'information' which the environment 'affords' the organism. As an ecological theorist, however, Gibson recognizes the importance of the organism in determining what is afforded. He details especially the role of movement and the temporal organization of the organism-environment relationship that results. Still, that organization does *not* consist of the construction of percepts from their elements; rather the process is one of responding to the invariances in that relationship. Thus perceptual learning involves progressive differentiation of such invariances, not the association of sensory elements.

The problem for me has been that I agree with all of the positive contributions to conceptualization which Gibson has made, yet find myself in disagreement with his negative views (such as on 'images') and his ultimate philosophical position. If indeed the organism plays such a major role in the theory of ecological perception, does not this entail a constructional position? Gibson's answer is no, but perhaps this is due to the fact that he (in company with so many other psychologists) is basically uninterested in what goes on inside the organism.

What, then, does go on in the perceptual systems that is relevant to this argument? I believe that to answer this question we need to analyze what is ordinarily meant by 'image'. Different disciplines have very different definitions of this term.

The situation is similar to that which obtained in neurology for almost a century with regard to the representation we call 'motor'. In that instance the issue was stated in terms of whether the representation in the motor cortex was punctile or whether in fact movements were represented. A great number of experiments were done. Many of them using anatomical and discrete electrical stimulation techniques showed an exquisitely detailed anatomical mapping between cortical points and muscles and even parts of muscles (Chang, Ruch and Ward, 1947). The well known homunculus issued from such studies on man (Penfield and Boldrey, 1937).

But other, more physiologically oriented experiments provided different results. In these it was shown that the same electrical stimulation at the same cortical locus would produce *different* movements depending on such other factors as position of the limb, the density of stimulation, the state of the organism (e.g. his respiratory rate, etc.). For the most part, one could conceptualize the results as showing that the cortical representation consisted of movement centered on one or another joint (e.g., Phillips, 1965). The controversy was thus engaged — proponents of punctate muscle representation vis à vis the proponents of the representation of movement.

I decided to repeat some of the classical experiments in order to see for myself which view to espouse (reviewed in Pribram, 1971, Chapters 12 and 13). Among the experiments performed was one in which the motor cortex was removed (unilaterally and bilaterally) in monkeys who had been trained to open a rather complex latch box to obtain a peanut reward (Pribram *et al.*, 1955—56). My results in this experiment were, as in all others, the replication of the findings of my predecessors. The latch box was opened, but with considerable clumsiness, thus prolonging the time taken some two- to three-fold.

But the interesting part of the study consisted in taking cinematographic pictures of the monkey's hands while performing the latch-box task and in their daily movements about the cage. Showing these films in slow motion we were able to establish to our satisfaction that no movement or even sequence of movements was specifically impaired by the motor cortex resections! The deficit appeared to be *task* specific, not muscle or movement specific.

My conclusion was therefore that depending on the *level of analysis*, one could speak of the motor representation in the cortex in three ways. Anatomically, the representation was punctate and of *muscles*. Physiologically, the representation consisted of mapping the muscle representation into *movements*, most likely around joints as anchor points. But behavioral analysis showed that these views of the representation were incomplete. No muscles were paralyzed, no movements precluded by total resection of the representation. *Action*, defined as the environmental consequence of movements, was what suffered when motor cortex was removed.

The realization that acts, not just movements or muscles were represented in the motor systems of the brain accounted for the persistent puzzle of motor equivalences. We all know that we can, though perhaps clumsily, write with our left hands, our teeth, or if necessary, our toes. These muscle systems may never have been exercised to perform such tasks, yet immediately and without practice can accomplish at least the rudiment required. In a similar fashion, birds will build nests from a variety of materials, and the resulting structure is always a habitable facsimile of a nest.

The problem immediately arose of course as to the precise nature of a representation of an act. Obviously there is no 'image' of an action to be found in the brain if by 'image' one means specific words or the recognizable configuration of nests. Yet some sort of representation appears to be engaged that allows the generation of words and nests — an image of what is to be achieved, as it were.

The precise composition of images-of-achievement remained a puzzle for many years. The resolution of the problem came from experiments by Bernstein (1967) who made cinematographic records of people hammering nails and performing similar more or less repetitive acts. The films were taken against black backgrounds with the subjects dressed in black leotards. Only joints were made visible by placing white dots over them.

The resulting record was a continuous wave form. Bernstein performed a Fourier analysis on these wave forms and was invariably able to predict within a few centimeters the amplitude of the next in the series of movements.

The suggestion from Bernstein's analysis is that a Fourier analysis of the invariant components of motor patterns (and their change over time) is computable and that an image-of-achievement may consist of such computation. Electrophysiological data from unit recordings obtained from the motor cortex have provided preliminary evidence that, in fact, such computations are performed (Evarts, 1967, 1968).

By 'motor image' therefore we mean a punctate muscle-brain connectivity that is mapped into movements over joints in order to process environmental invariants generated by or resulting from those movements. This three-level definition of the motor representation can be helpful in resolving the problems that have become associated with the term 'image' in perceptual systems.

In vision, audition and somesthesis (and perhaps to some extent in the chemical senses as well) there is a punctate connectivity between receptor surface and cortical representation. This anatomical relationship serves as an *array* over which sensory signals are relayed. At a physiological level of analysis, however, a mapping of the punctate elements of the array into functions occurs. This is accomplished in part by convergences and divergences of pathways but even more powerfully by networks of lateral interconnectivities, most of which operate by way of slow graded dendritic potentials rather than by nerve impulses propagated in long axons. Thus in the retina, for instance, no nerve impulses can be recorded from receptors, bipolar or horizontal cells. It is only in the ganglion cell layer, the last stage of retinal processing, that nerve impulses are generated to be conducted in the optic nerve to the brain (reviewed in Pribram, 1971, Chapters 1, 6, and 8). These lateral networks of neurons operating by means of slow graded potentials thus map the punctate receptor-brain connectivities into functional *ambiences*.

By analogy to the motor system, this characterization of the perceptual process is incomplete. Behavioral analysis discerns perceptual constancies just as it had to account for motor equivalences. In short, *invariances* are processed over time and these invariances constitute the behaviorally derived aspects of the representation (see e.g. Pribram, 1974b). Ordinarily, an organism's representational processes are called *images* and there is no good reason not to use this term. But it must be clearly kept in mind that the perceptual image, just as the motor image, is more akin to a computation than to a photograph.

We have already presented the evidence that for the visual system at least, this computation (just as in the motor system) is most readily accomplished in the Fourier or some similar domain. The evidence that pattern perception depends on the processing of spatial frequencies has been reviewed. It is, after all, this evidence more than any other that has suggested the holonomic hypothesis of perception.

The perceptual image, so defined, is therefore a representation, a mechanism based on the precise antomical punctate receptor-cortical connectivity that composes an *array*. This array is operated upon by lateral interconnections that provide the *ambiences* which process the *invariances* in the organism's input. The cortical representation of the percepts go therefore beyond the anatomical representations of the receptor surfaces just as the cortical representation of actions goes beyond the mere anatomical representations of muscles.

It is, of course, a well known tenet of Gestalt psychology that the percept is not an equivalent of the retinal (or other receptor) image. This tenet is based on the facts of constancy (e.g. size) and the observations of illusions. Neurophysiologists, however, have only recently begun to seriously investigate this problem. Thus Horn (Horn *et al.*, 1972) showed that certain cells in the brainstem (superior colliculus) maintained their firing pattern to an environmental stimulus despite changes in body orientation; and in my laboratory Spinelli (1970) and also Bridgeman (1972) using somewhat different techniques demonstrated constancy in the firing pattern of cortical neurons over a range of body and environmental manipulations. Further, neurobehavioral studies have shown that size constancy is impaired when perivisual and inferior temporal cortex is removed (Humphrey and Weiskrantz, 1969; Ungerleider, Gauz and Pribram, 1977).

The fact that the cortex becomes tuned to environmental invariances rather than just to the retinal image is borne out dramatically by a hitherto unexplained discrepancy in the results of two experiments. In both experiments a successful attempt was made to modify the orientation selectivity of the cortical neurons of cats by raising them from birth in environments restricted to either horizontal or vertical stripes. In one experiment (Blakemore, 1974) the kittens were raised in a large cylinder appropriately striped. A collar prevented the animals from seeing parts of their bodies – so they were exposed to only the stripes. However, and this turns out to be critical, the kittens could observe the stripes from a variety of head and eye positions. By contrast, in the other experiment, which was performed in my laboratory (Hirsch and Spinelli, 1970), head and eye turning was prevented from influencing the experiment by tightly fitting goggles onto which the stripes were painted. In both experiments cortical neurons were found to be predominantly tuned to the horizontal or vertical depending on the kitten's environment, although the tuning in Blakemore's experiments appeared to be somewhat more effective. The discrepancy arose when behavioral testing was instituted. Blakemore's kittens were consistently and completely deficient in their ability to follow a bar moving perpendicular to the orientation of the horizontally or vertically striped environment in which they had been raised. In our experiment Hirsch, despite years of effort using a great number of quantitative tests, could never demonstrate *any* change in visual behavior! The tuning of the cortical cells to the environmental situation which remained invariant across transformations of head and eye turning was behaviorally effective; the tuning of cortical cells to consistent retinal stimulation had no behavioral consequences.

These results are consonant with others obtained in other sensory modes and also help to provide some understanding of how brain processing achieves our perception of an objective world separated from the receptor surfaces which interface the organism with his environment.

Von Bekesy (1967) has performed a large series of experiments on both auditory and somatosensory perceptions to clarify the conditions that produce projection and other perceptual effects. For example, he has shown that a series of vibrators placed on the forearm will produce a point perception when the phases of the vibrations are appropriately adjusted. Once again, in our laboratory we found that the cortical response to the type of somatosensory stimulation used by Bekesy was consonant with the perception, not with the pattern of physical stimulation of the receptor surface (Dewson, 1964; Lynch, 1971). Further, Bekesy showed that when such vibrators are applied to both forearms, and the subject wears them for a while, the point perception suddenly leaps into the space between the arms.

Other evidence for projection comes from the clinic. An amputated leg can still be perceived as a phantom for years after it has been severed and pickled in a pathologist's jar. A more ordinary experience comes daily to artisans and surgeons who 'feel' the environment at the ends of their tools and instruments.

These observations suggest that direct perception is a special case of a more universal experience. When what we perceive is validated through other senses or other knowledge (accumulated over time in a variety of ways, e.g., through linguistic communication — see Gregory, 1966), we claim that perception to be veridical. When validation is lacking or incomplete, we tend to call the perception an illusion and pursue a search for what physical events may be responsible for the illusion. Gibson and his followers are correct, perception is direct. They are wrong if and when they think that this means that a constructional brain process is ruled out or that the percept invariably and directly gives evidence of the physical organization that gives rise to perception.

As noted, there is altogether too much evidence in support of a brain constructional theory of perception. The holonomic model, because of its inclusion of parallel processing and wave interference characteristics, readily handles the data of projection and illusion that make up the evidence for direct perception. The holonomic model also accounts for the 'directness' of the perception: holographic images are not located at the holographic plane, but in front or beyond it, away from the constructional apparatus and more into the apparently 'real', consensually validatable external world.

7. THE NATURE OF THE PERCEIVED UNIVERSE

In the concluding part of this paper, I want, therefore, to explore some questions as to the organization of this external 'real' physical world. Unless we know something of consensually validatable 'information' that remains invariant across transformations of the input to the brain — and, as we have seen, we cannot rely only on the directness of our perceptual experience for this knowledge — how can we think clearly about what is being perceived? Questions as to the nature of the physical universe lie in the domain of the theoretical physicist. Physics has enjoyed unprecedented successes not only in this century, but in the several preceeding ones. Physics ought to know something, therefore, about the universe we perceive. And, of course, it does. However, as we shall shortly see, the structure-distribution problem is as pervasive here as it is in brain function.

The special theory of relativity made it clear that physical laws as conceived in classical mechanics hold only in certain circumscribed contexts. Perceptions of the Brownian 'random' movements of small suspended particles, or of the paths of light coming from distances beyond the solar system strained the classical conceptions to the point where additional concepts applying to a wider range of contexts had to be brought in. As in the case of direct perception, the laws of physics must take into account not only what is perceived but the more extended domain in which the perception occurs. The apparent flatness of the earth we now know as an illusion.

The limitations of classical physics were underscored by research into the microcosm of the atom. The very instruments of perception and even scientific observation itself became suspect as providing only limited, situation related information. Discrepancies appeared such as an electron being in two places (orbits) at once or at best moving from one place to another faster than the speed of light — the agreed upon maximum velocity of any event. And within the nucleus of the atom matters are worse — a nuclear particle appears to arrive in one location before it has left another. Most of these discrepancies result from the assumption that these particles occupy only a point in space — thus when the equations that relate location to mass or velocity are solved, they lead to infinities. Furthermore, in the atomic universe, happenings take place in jumps – they appear to be quantized, i.e., particulate. Yet when a small particle such as an electron, or a photon of light, passes through a grating and another particle passes through a neighboring grating, the two particles appear to interact as if they were waves, since interference patterns can be recorded on the far side of the gratings. It all depends on the situation in which measurements are made whether the 'wavicle' shows its particle or its wave characteristics.

Several approaches to this dilemma of situational specificity have been forwarded. The most popular, known as the Copenhagen solution, suggests that the wave equations (e.g., those of Schrödinger, 1935; and deBroglie, 1964) describe the average probabilities of chance occurrences of particulate events. An earlier solution by Niels Bohr (the 'father' of the Copenhagen group, 1966) suggested that particle and wave were irreconcilable complimentary aspects of the whole. Heisenberg (1959) extended this suggestion by pointing out that the whole cannot in fact be known because our knowledge is always dependent on the experimental situation in which the observations are made. Von Neumann (1932) added, that given a positivistic operational framework, the whole reality becomes therefore not only unknown but unknowable. Thus the whole becomes indeterminable because we cannot in any specific situation be certain that what we are observing and measuring reflects 'reality'. In this sense, as well as from the viewpoint of brain processes, we are always constructing physical reality. The arguments of the quantum physicist and those of the neurophysiologist and psychologist of perception are in this respect identical.

But several theoretical physicists are not satisfied with these solutions or lack of solutions. Feynman (1965), for instance, notes that though we have available most precise and quantitative mathematical descriptions in quantum mechanics, we lack good images of what is taking place. (His own famous diagrams show time flowing backwards in some segments!) DeBroglie, who first proposed wavelike characteristics for the electron fails to find solace in a probabilistic explanation of the experimental results that led him to make the proposal (1964). And DeBroglie is joined by Schrödinger (1935) who formulated the wave equation in question and especially by Einstein, whose insights led him to remain unconvinced that an unknowable universe, macro- and micro-, was built on the principle of the roulette wheel or the throw of dice.

I share this discomfort with attributing too much to chance because of an experience of my own. In the Museum of Science and Industry in

Chicago, there is a display which demonstrates the composition of a Gaussian probability distribution. Large lead balls are let fall from a tube into an open maze made of a lattice of shelves. The written and auditory explanations of the display emphasize the indeterminate nature of the path of each of the falling balls and provide an excellent introduction to elementary statistics. However, nowhere is mention made of the symmetrical maze through which the balls must fall in order to achieve their probabilistic ending. Having just completed *Plans and the Structure of Behavior* (Miller et al., 1960), I was struck by the omission. In fact, students of biology routinely use statistics to discover the orderliness in the processes they are studying. For example, when a measurable entity shows a Gaussian distribution in a population, we immediately look for its heritability. Perhaps the gas laws from which statistics emerged have misled us. A Gaussian distribution reflects symmetrical *structure* and not just the random banging about of particles. Again, the physical reality behind the direct perception may contain surprises.

Moreover, when we obtain a probabilistic curve, we often refer to a distribution of events across a population of such events — e.g., a Gaussian distribution. Could it be that for the physical universe, just as in the case of brain function, structure and distribution mutually interact? After all, the brain is a part of the physical universe. For brain function, we found structure to be in the form of program and distribution in the form of holograms. Is the rest of the physical universe built along these lines as well?

David Bohm (1957), initially working with Einstein, has among others, made some substantial contributions to theoretical physics compatible with this line of reasoning. Bohm points out, as noted above, that the oddities of quantum mechanics derive almost exclusively from the assumption that the particles in question occupy only a point in space. He assumes instead that the "wavicle" occupies a finite space which is structured by subquantal forces akin to electromagnetic and gravitational interactions. These interacting forces display fluctuations — some are linear and account for the wave form characteristics of the space or field. Other interactions are nonlinear (similar to turbulance in fluid systems) and on occasion produce quantal events. In biology, Thom (1972) has developed a mathematics to deal with such occurrences in the morphogenetic field and this mathematics has been applied to perception by Bruter (1974). Thom calls the emergence of quasistable structures from turbulent processes 'catastrophes'. In physics, the quantal structures that result from such catastrophic processes may, therefore, be only partially stable. Thus, they can disappear and reappear nearby in a seemingly random fashion; on the average, however, they would be subject to the more regular oscillations of the subquantal forces. In biology, observations pertaining to the entrainment of oscillatory processes by clocks or temporary dominant foci parallel these concepts. Bohm goes on to point out where in the subquantal domain these events will become manifest: the interactions of high frequency and high energy particles in nuclear reactions, in black bodies, etc. An article in a recent issue of *Scientific American* reviews the contemporary scene in these attempts at a Unified Field Theory in the subquantal domain (Weinberg, 1974).

More recently, Bohm (1971, 1973) has reviewed the conceptual development of physics from Aristotelian through Gallilean and Newtonian times to modern developments in the Quantum Mechanics. He points out how much of our image of the physical universe results from the fact that, since Galileo, the opening of new worlds of inquiry in Physics has depended on the use of lenses. Lenses have shaped our images and lenses objectify. Thus we tend to assess external space in terms of objects, things and particulars.

Bohm goes on to suggest that image formation is only one result of optical information processing and proposes that we seriously consider the hologram as providing an additional model for viewing the organization of physical processes. He and his group are now engaged in detailed application of this basic insight to see whether in fact a holographic approach can be helpful in solving the problems of high energy nuclear physics. Initial developments have shown promise.

As noted above, the subquantal domain shows striking similarities to holographic organization. Just as in the case for brain processes presented here, Bohm's theoretical formulations retain classical and quantum processes as well as adding the holographic. The holographic state described by wave equations and the particle state described quantally, are part of a more encompassing whole. The parallel holds because the holographic models describe only the deeper levels of the theory which is thus holonomic, rather than holographic, as we found it to be for the special case of brain function (where the deeper level is constituted of pre and post synaptic and dendritic potentials and the quantal level, of the nerve impulses generated by these slow potentials).

Bohm relates structural and holographic processes by specifying the differences in their organization. He terms classical and particle organization *explicate* and holographic organization *implicate*. Elsewhere (Pribram, in press), I have made a parallel distinction for perceptual processes: following

Bertrand Russell (1959), I proposed that scientific analysis as we practice it today, begets knowledge of the extrinsic properties (the rules, structures, etc.) of the physical world. My proposal departs from Russell, however, in suggesting that intrinsic properties (which he defines as the stoneness of stones, e.g.) are also knowable — that in fact they are the 'ground' in which the extrinsic properties are embedded in order to become realized. Thus artists, artisans and engineers spend most of their time realizing the extrinsic programs, laws and rules of the arts and sciences by grounding them in an appropriate medium. For example, a Brahms symphony can be realized by an orchestra, on sheet music, on a long-playing record, or on tape. Each of these realizations come about after long hours of development of the medium in which the realization occurs. Russell was almost correct in his view that the intrinsic properties of the physical world are unknowable — they have apparently little to do with the more enduring extrinsic properties, show no resemblances amongst themselves, and demand considerable know-*how* to replicate.

The sum of these ideas leads to the proposal that the intrinsic properties of the physical universe, their implicate organization, the field, ground or medium in which explicit organizations, extrinsic properties become realized, are multiform. In the extreme, the intrinsic properties, the implicate organization, is holographic. As extrinsic properties become realized, they make the implicate organization become more explicit.

The consequence for this view is a re-evaluation of what we mean by probabilistic. Until now, the image, the model of statistics, has been indeterminacy. If the above line of reasoning is correct, an alternate view would hold that a random distribution is based on holographic principles and is therefore determined. The uncertainty of occurrence of events is only superificial and is the result of holographic 'blurring' which reflects underlying symmetries (much as does the Gaussian distribution in our earlier example) and not just haphazard occurrences. This relation between appearance and reality in the subquantal domain of nuclear physics and its dependence on underlying symmetries (spin) is detailed in the review article in *Scientific American* already referred (Weinberg, 1974).

A preliminary answer to the question posed at the outset of this section — what is it that we perceive — is therefore that we perceive a physical universe not much different in basic organization from that of the brain. This is comforting since the brain is part of the physical universe as well as the organ of perception. It is also comforting to find that the theoretical physicist working from his end and with his tools and data has come to the identical problem (which is, in Gibson's terms, the nature of the information which remains invariant across situations) faced by the neurophysiologist and psychologist interested in perception (Bohm, 1965, Appendix). Though surprising, the fact that at least one renowned theoretical physicist has made a proposal that addresses this common problem in terms similar to those set forth on the basis of an analysis of brain function, is most encouraging. For science is of a piece, and full understanding cannot be restricted to the developments made possible by one discipline alone. This is especially true for perception — where perceiver meets the perceived and the perceived meets the perceiver.

Department of Psychology, Stanford University Stanford, California

NOTE

* This work was supported by NIMH Grant No. MH12970-08 and NIMH Career Award No. MH15214-13 to the author.

MARY HENLE

ON THE DISTINCTION BETWEEN THE PHENOMENAL AND THE PHYSICAL OBJECT

My assignment is to discuss the distinction between the phenomenal and the physical object. The failure to make this distinction is the doctrine of naive realism, a position that is as persistent as it is wrong. I shall start with Bertrand Russell's statement of the case against naive realism:

Naive realism leads to physics, and physics, if true, shows that naive realism is false. Therefore naive realism, if true, is false; therefore it is false.[1]

By this remark I assume that Russell means that the identification of the perceptual world with the physical world led to the investigation of this physical world. The apparent objectivity of percepts invited scientific study of the world of objects. But then optics showed that the physical object is lost in the chain of events that leads to the perception of an object. From the physical object, whose existence nobody challenges, are reflected light beams of varying frequencies and intensities; these reach the eye from different directions. The important point is that the light beams are independent of one another. Thus an array of independent beams reaches the eye. While the phenomenal object is a unity, the corresponding optic array is an arrangement of independent beams. This discovery meant that it is impossible to identify the physical with the phenomenal object — that would be to identify an arrangement with a unity, an aggregate with an organization.

Some of the light beams, as we know, stimulate receptors in the retina, and from here on up to the cerebral cortex, interactions take place which produce the unity, the organization corresponding to the phenomenal object. But the percept is the end product of a series of events which only *starts* with the physical object.

Now it must be admitted that in private life each of us is a naive realist — until the identification of phenomenal and physical object is called into question, as in discussions such as this one. And yet this sturdy philosophy begs important questions of perception — most conspicuously the problem of organization, as I have just indicated. Thus it would seem to be tenable only if one takes perception for granted, as we do in everyday

John M. Nicholas (ed.), Images, Perception, and Knowledge, 187–193.
All Rights Reserved.
Copyright © 1977 by D. Reidel Publishing Company, Dordrecht-Holland.

life. Then there is nothing against naive realism — except physics. And since our naive physics is no more sophisticated than our naive philosophy, it does not constitute a formidable objection.

Many behaviorists, especially but not exclusively the earlier behaviorists, likewise hold implicitly to a doctrine of naive realism because they take perception for granted. But for behaviorism this has the unfortunate consequence of creating an intolerable ambiguity in the definition of the key term *stimulus*. Does stimulus refer to physiological processes, for example in retinal cells — to proximal stimulation? More often it means the 'stimulus object' that elicits a response; but is this object physical or phenomenal? It is impossible to tell if the two are not distinguished. But it is important to know because, regardless of the intentions of the psychologist, only if the phenomenal object is meant is it intelligible that it elicit a response.

There are other authors who do not take perception for granted — who are interested in the nature of perception — and who nevertheless fail to distinguish between phenomenal and physical object. This is the more interesting case, and I should like to look into it.

It seems to me that the motive behind naive realism in this more interesting case is the fear of subjectivity. If we make the distinction between the objects of our experience and physical objects, we acknowledge that physical objects are not directly accessible to us. Then we live each in our own world, knowing only our perceptions, cut off from the objective environment as well as from the world of others. How can we then make the observations necessary for the development of an objective science? How can we hope to confirm the observations of others if each of us is confined to his own phenomenal world? How can we understand other people? Have we not painted ourselves into the ridiculous corner of solipsism by the rejection of naive realism?

The question then comes down to that of whether the distinction between phenomenal and physical object leads to subjectivity. May I repeat that, whether we like it or not, physical and phenomenal objects *are* separated by a sequence of events in the physical world and in the nervous system in which the unified object is lost. And may I add that if the psychologist is confined to his phenomenal world, so too is the physicist. But the physicist, unconcerned with his epistemological predicament, proceeds to report perceptual data — pointer readings and the like — on the basis of which he constructs his science. Psychologists do not object that physics is subjective, even though it depends for its data on the phenomenal fields of observers in exactly the same way that psychology does. It would seem that, whether or not the phenomenal field is subjective, it does not prevent the development of science.

But is the phenomenal field subjective? It is time to examine some of the meanings of this elusive term. Köhler has pointed out that we need to distinguish between at least two meanings of subjectivity.[2] All experiences are subjective in the sense that they depend upon processes in the physiological organism: perceptions, images thoughts, emotions, etc. are equally subjective in this sense — and they are subjective in origin whether the physiological organism in question belongs to a physicist, a psychologist, or a layman. But it is misleading to leave the problem there. We need to recall the distinction between physical and phenomenal object — here specifically the distinction between the physiological (physical) organism and the phenomenal self. While all experiences are subjective in their *origin*, they are not all subjective in their *locus* or reference. Some experiences are localized in the phenomenal self, others in objects outside this phenomenal self. My perception of that blue book, for example, depends on processes in my visual system and is in that sense subjective — in origin it is subjective. But the form and the blueness are located in the phenomenal book outside my phenomenal self and are, in that sense, objective. My interest in the contents of the volume, on the other hand, is located in my phenomenal self, and is thus subjective, although this same self is perceived as directed to events outside itself.

In short, the assertion that the distinction between physical and phenomenal object leads to subjectivity depends in part on that last refuge of naive realism, the confusion of the physical organism with the phenomenal self. Once we make the distinction, we see that neither the pointer readings of the physicist nor the behavioral observations of the psychologist are subjective; they are objective since they refer to phenomenal objects outside the (phenomenal) self.

It has, of course, to be conceded that the phenomenal field is private, even though it may contain objective as well as subjective facts. I cannot share your perceptual field; I cannot know your thoughts, and when I try to guess them, the mental contents involved are mine, not yours. The privateness of experience has been a source of embarrassment to psychologists. Various solutions have been proposed. Naive realism essentially denies that there is a problem. Methodological behaviorism advises us to confine our observations to 'publicly observable behaviors of organisms'. But it is obvious that, when intersubjective agreement is achieved, it is achieved by observers each of whom is reporting facts in his own private phenomenal field — the only facts accessible to him. Intersubjective agreement thus does not overcome the privacy of experience. Another solution has been in terms of increasing privacy of perception. I quote:

> All observers can obtain exactly the same information about a tree if they all walk around it and get the same perspectives. Each observer gets a somewhat different set of perspectives of his own hands than any other observer gets, although there is much in common. But the perspective of one's own nose is absolutely unique and no one else can ever see it from that particular point of view. . . . The tree, the hand, the nose are increasingly private. [3]

In these examples, differences indeed exist between the perspectives of different observers and thus in the extent to which anyone's particular description can be confirmed. These are differences in the likelihood of intersubjective agreement; but once more, each observer remains confined to his own experience: the *phenomenal* tree, hand, and nose are equally private.

In view of the inevitable privacy of experience, I prefer, with Köhler, to make a virtue of necessity:

> When the behaviorist says that the private 'subjective' content of the phenomenal world cannot interest a true scientist, this statement is plainly contradicted by his own procedure. He actually proves that certain contents of the directly accessible world must, and can, be used as reliable tools in his science. . . . [Perceptual] scenes are accessible to only one person in each case? This may be true; but then, they are accessible to at least one, while the independently existing objects in the sense of physics, including the physical behavior of physical men and animals, are directly accessible to *nobody*.[4]

May I summarize to this point: to deny the distinction between phenomenal and physical objects is to overlook the processes that are responsible for perception. It is a position that arises either from taking perception for granted or from fear of subjectivity. Thus various meanings of subjectivity and objectivity were examined. Even the privacy of phenomenal experience is no obstacle to the development of science; rather, the material directly accessible to the observer provides him with all the scientific data he can obtain. There seems to be no scientific justification for failing to distinguish between phenomenal and physical facts.

Now I would like to go farther and suggest that the identification of phenomenal and physical object actually impoverishes our treatment of perception. A. J. Ayer has remarked:

A philosopher who thinks that he directly perceives physical objects does not for that reason expect anything different to happen from what is expected by one who believes that he directly perceives sense-data.[5]

For that last phrase we need, of course, to substitute a phrase like 'one who believes that he directly perceives the phenomenal world'. No philosopher directly perceives sense data. But even though they might not expect anything different to happen, the two philosophers are likely to discuss perception differently. If we believe that we perceive physical objects, we are likely to limit our treatment of perception as much as possible to the austere categories of physics. Then we will admit as 'real' perceptual facts only such material as the physicist might use in his observations, or at any rate we are likely to favor such facts. The result will be that we will exclude Gestalt or Ehrenfels or tertiary qualities.[6]

Does the drooping willow look sad? Nonsense – it is only our sadness that we see projected onto the willow. Does a house look friendly? Nothing of the sort: it is our own good feeling that we see in the inanimate object. What does a house know about friendliness? is a melody lively? a gait clumsy? a gesture menacing? Again, these are but the products of our own higher mental processes. If a meal looks inviting or a uniform forbidding – once more, these value experiences are something we ourselves add to the perceptual data.

But how does our sadness know to attach itself to the drooping willow and not to the floating dogwood? Why does our friendliness select one house rather than another as its object? Why is a melody lively even when our own state of mind is quite different? In all these instances we are overlooking the expressive characteristics of the percept itself and attributing them to properties of the self. And – within such a framework – we are ignoring the question of why a property of the self attaches itself to one particular percept and not to another. A friendly house may make us feel welcome, but it must first look friendly to do so. If the expressive character is in the percept, it is not necessary – and often not plausible – to add an explanation in terms of a projected state of mind. Rudolf Arnheim has gone far in showing the formal similarities between particular perceptions and the emotional and other characteristics they express.[7] We perceive the sadness of the willow as truly as we perceive its yellow green color and the shape of its flexible branches.

Once we start adding auxiliary assumptions to account for perceptual properties, we can go in a number of directions, all equally arbitrary with respect to particular percepts – unconscious inferences, influences of past experience, and so on. I would like to raise certain questions that apply to all such assumptions. How was the percept experienced the first time – before past experience or judgment or projection could act upon it? If a perception is seen as organized in a particular way the first time, without benefit of, say, past experience, why is experience needed at all to account for it? The same question has just been asked about the expressive characteristics of perception. Again, how is the experience itself (or the projection) aroused, if not by a percept already organized in a particular way? Once this organization exists, a perception can evoke all sorts of meanings and other influences of past experience; but it requires a particular organized percept to make contact with memory traces, themselves presumably aftereffects of previous perceptual organizations. Again, *how* does judgment or experience or projection operate to transform our perceptions? Such questions are usually not raised, but they are crucial if these factors are to be given an explanatory role in perception. How do we know that such processes occur? By definition, we are not aware of unconscious inferences, and the other processes mentioned are likewise immune to observation. We assume their operation because the percept does not show the characteristics we expected it to show. To salvage our expectation, we thus posit an untestable process. This is not the way that science ordinarily proceeds: scientific observation usually takes precedence over the investigator's expectations. And, finally, what kind of theory requires the auxiliary assumptions of which I am speaking? I am suggesting that one kind is a theory impoverished by neglecting or minimizing those aspects of perception which do not figure in the physicist's observations. This is a sensation-based theory of perception.

And this brings us back to the distinction between phenomenal and physical object. If we distinguish clearly between the two, we are not forced to adopt these auxiliary assumptions. It is ironic to note that these assumptions are stated in terms of subjective influences, influences that refer to the self. Thus, by the failure to consider the phenomenal object in its own right, on its own terms, we are forced into subjectivity. This is the very trap which those who fail to distinguish between phenomenal and physical object seek to avoid.

New School for Social Research

NOTES

[1] Russell (1940), p. 15.
[2] See, for example, Köhler (1947), Chapter 1, and Köhler (1938), pp. 69–70.
[3] Gibson (1967), p. 171.
[4] See Köhler in Feyerabend and Maxwell (1966), p. 86.
[5] Ayer (1956), p. 85.
[6] For example, see Köhler (1947), Chapter 6; Köhler (1938), pp. 78–79; and Köhler (1937).
[7] Arnheim (1949).

D. W. HAMLYN

UNCONSCIOUS INFERENCE AND JUDGMENT IN PERCEPTION

The notion of unconscious inference and the suggestion that it has a part to play in perception is, I suppose, mainly associated with the name of Helmholtz. He used it to explain those cases where the way in which we perceive things deviates from what would be expected on the basis of the pattern of stimulation alone — in connection, for example, with the phenomena of colour contrast, or size perception. Though I put it in this way — in terms of the pattern of stimulation — this is in fact not quite correct. For, why should not the phenomena of colour contrast, for example, be explained entirely in terms of the pattern of stimulation on the retina, or if not there in the cortex, as long as interaction between areas of excitation is allowed? Why should it not be the case that a grey patch looks different when set alongside or in the context of an area of red from what it does when set in a neutral context? And what prevents our explaining this, at any rate partially, by reference to the effects that the stimulation of one area of the retina has on the effects in another area produced by accompanying or parallel stimulation, wherever in the nervous system the final integrating mechanism is to be found? (I make this last qualification for reasons that will appear directly.)

Helmholtz's need to invoke another factor in such cases must therefore be due to his embracing a theory which ruled out such explanations as I have mentioned. And this is indeed the case. For he embraced what has become best known through the criticisms of the Gestalt Psychologists as the 'constancy hypothesis' — that there should, unless other factors intervene, always be a constant relation between how we perceive things and the pattern of stimulation. I suspect that the constancy hypothesis has at its back another doctrine — that of 'specific nervous energies' — that Johannes Müller put forward in the 1830s, although he himself acknowledged its source in another physiologist named Bell. On this theory each nerve has its own function and cannot take over another function. It was inferred from this that the excitation of any given nerve attached to a sense-organ produces its own experience. Such a theory is naturally associated with and could be used to bolster up, firstly, a sensory atomism — the thesis that in

John M. Nicholas (ed.), Images, Perception, and Knowledge, 195–212.
All Rights Reserved.
Copyright © 1977 by Reidel Publishing Company, Dordrecht-Holland.

vision, for example, excitation of the eye produces a mosaic of discrete and atomic visual experiences. Secondly, it could be used to provide the backing of the constancy hypothesis to which I have already referred, since the hypothesis of discrete neural functioning naturally has as a corollary the thesis that what takes place at the level of nerve-endings must be repeated at the level of resulting experience. And if the neural functioning at the level of nerve-endings is discrete so must be the array of experiences which result. Discrete functioning rules out interactive processes. One can see from this why the Gestaltists, who were so opposed to the idea of discrete functioning in the nervous system, were equally opposed to the constancy hypothesis. But the thesis of isomorphism — that there is an identity or complete similarity of structure between the phenomenal (how things appear) and the cortical (how things are in the brain) — is nevertheless a theory in the same line of country. For it assumes that there is a correspondence between how things appear and the pattern of neural excitation, even if it rejects the idea that the correspondence must be at the level of what happens at nerve-endings.

Helmholtz thought that the deviations of perceptual experience from what is to be expected on the basis of the constancy hypothesis are to be explained not in terms of anything further in the process of stimulation of sense-organs (for how could it be on the assumptions in question?) but in terms of 'intellectual processes' carried out by the perceiver. In particular the perceiver makes some kind of judgment or carries out some kind of inference, which, since he is not aware of it, must therefore be unconscious. The role that this might play is most obvious in the case of space-perception where the sizes and shapes that things seem to have often deviate from what might be expected from a consideration of the pattern of stimulation on the retina alone. Helmholtz also used the idea in connection with such phenomena as colour contrast. The Helmholtz three-cone theory of colour vision allows a deviation from the thesis of one experience — one kind of nerve being stimulated, in that it allows that some experiences, the perception of yellow, for example, may be the product of the excitation of more than one kind of cone simultaneously in a given proportion. But it does not allow for interactive processes between areas of the retina stimulated in the ordinary way. That is to say that a set of cones being stimulated by light of a certain frequency should produce a definite experience whatever be going on in adjacent areas of the retina. But this does not seem to be the case. Hence it has to be explained, given these presuppositions, in a quite different way. It is assumed that a grey patch seen in the middle of a red area will tend to look green because the perceiver makes an inference about the colour of the inset patch. That is to say that there is a tendency to infer that the inset colour must be opposed to the colour of the surrounding area, and this inference sometimes overrides the given facts of sensory experience. Helmholtz thought that it does this because we have the tendency to take the inset patch as soon *through* the surrounding colour.

My phrase 'the given facts of sensory experience' is important because it reveals that there is presupposed in all this a thesis about the relationship of sensation to perception. The constancy hypothesis in fact assumes that, at least over the range of phenomena where it applies, perception is just a matter of sensation. Thus where the hypothesis cannot get application the deviations that are apparent cannot be due to anything sensory and must therefore be explained by reference to processes that are not sensory.[1] In fact the rejection of the constancy hypothesis by the Gestalt psychologists and its replacement by the thesis of isomorphism is despite first appearances a kind of reinstatement of the role of the sensory in perception, except that it is recognised that the nature and structure of the sensory is not determined simply by what happens at the level of nerve endings. At all events it rejects the idea that anything non-sensory or intellectual plays a part in perception, the gap between what might be expected on the basis of what happens at the level of nerve-endings and what appears to be the case from a consideration of the phenomenal being closed by appeal simply to further or different causal physiological processes. It is perhaps of some interest in this connection that J. J. Gibson claimed in his *The Perception of the Visual World* to be resuscitating the constancy hypothesis by providing a more comprehensive account of what must be taken as happening at the level of the retina; he thereby ruled out any role for anything like unconscious inference. He also claimed that his theory was a psychophysical one. In his more recent book *The Senses Considered as Perceptual Systems*, which in many ways extends the account of the earlier book, he makes an explicit rejection of the notion of sensation as having any part to play in a theory of perception, as well as a rejection of any intellectual processes such as those involved in conception and belief. Yet in most ways the theory of the later book is just an extension of that of the earlier. How can he in that case reject the notion of the sensory in an account of perception?

The answer to this question lies, I think, in the tradition of thinking against which the psychology of perception has developed. For while it is natural perhaps to think that the stimulation of a sensory organ must eventually result in a sensation or group of sensations, it is not so clear exactly what a sensation is; and for that there is a whole epistemological tradition to fall back on. When I say that it is not clear exactly what a sensation is, I mean that this is not clear in the case of the majority of the senses. The suggestion that the stimulation of one kind of nerve ending may produce a sensation of pain is relatively unproblematic. But what counts as analogous to this in the case of, say, vision? The empiricist tradition in epistemology has generally had recourse to the notion of sense-impressions or sense-data as what is given in perception. It is thus natural, once given this notion, to suppose that these sense-impressions are what are caused by the stimulation of our senses — and the term 'sense-impression' is such that its very etymology supports that idea. Any suggestion that what we perceive and how we perceive it must be a function of construction or other activity on our part correspondingly undermines the view that perception is ultimately a matter of things being given via sense-impressions. And that is how it is with Gibson. I say nothing here about the rightness or wrongness of his view; I wish merely to make clear its ancestry and general character.

I have tried elsewhere, particularly in my *Sensation and Perception,* to make clear the role of sensation within perception. It may be enough now to say that what I have described above as natural — that it is sense-impressions that are the result of the direct stimulation of our sense-organs — is, when one comes to think about it, not natural at all. The supposal that it *is* comes from the pervasiveness of a tradition, and the way in which the psychology of perception has emerged out of a long course of philosophizing about perception with epistemological problems in mind. Though I find it an increasingly unpopular view these days, I am still of the opinion that distinguishing philosophical from psychological questions is a desirable prolegomenon to any fruitful thinking in this and other areas. Progress does not always come from everybody trying to tackle the same issues in the same way, without distinction of function, and without, to mention a more important point, getting clear what the problems are. Co-operation between philosophers and psychologists may be more fruitful if it comes from delimitation of function rather than an attempt to answer any question that presents itself without first sorting out what kind of question it is. However that may be, the question whether there is anything given in perception is the question whether the use of our senses provides us with any knowledge which is both incorrigible and independent of any knowledge of any other and prior kind. The empiricist who wishes to claim that knowledge is in the end totally dependent on the use of our senses must claim that there is knowledge of this kind. And such knowledge must not be concept-dependent in the sense that it presupposes, whether simply logically or also temporally, any conceptual understanding in terms of which what the senses provide is to be construed. For such conceptual understanding would involve a form of knowledge — a knowledge of how what is provided by the senses is to be construed, or, to put it in another way, what it is to be construed *as*. And in that case what the senses provide could not constitute a 'given'.

The issues here are about the foundations of knowledge and whether there are such foundations and if so where they are to be sought. The empiricist view that such foundations are to be sought in sense-impressions or sense-data does not necessarily have as a corollary that these must be *caused* by the stimulation of our senses, however natural it is to think that this must be the case. For one thing, there are alternative theories, however plausible or implausible they may seem, e.g. the view put forward by Berkeley that sensory ideas can only be caused by spirits. Such a view would be absolutely excluded if the thesis that sense-impressions are caused by the stimulation of our sense-organs was a necessary corollary of the thesis that sense-impressions provide the foundations of our knowledge. If it is natural to look in such a connection to the stimulation of our sense-organs, it is because a causal process of this kind seems to exclude the possibility of its results presupposing the application of knowledge derived from any other source. It remains however a question whether those results can be construed as knowledge at all, and if they cannot they are of no use in providing the foundations of knowledge. There would in that event be a gap between the causal story and the epistemological pursuits. In fact the ramification of these factors and the kinds of philosophical influences that have come to bear on psychological theories of perception have been very complicated; but basically the issues turn in the end on considerations of the kind that I have mentioned.

If we return to the constancy hypothesis, it is worth noting that what is supposed to be constant is the relation between the pattern of stimulation and the resulting pattern of sensations; yet those sensations must be thought to constitute forms of perception. For, anyone who embraces the constancy hypothesis ought to suppose that how we see the world can be explained ultimately in terms of perceptions corresponding to the sensations brought about by strictly sensory processes, and any other factors that are introduced simply build on the foundations of the original perceptions. The Helmholtzian notion of unconscious inference in a way makes this clear, for inference must take place from what is known or believed; hence if the inference is to be the explanation why we do not see things as the constancy hypothesis suggests it must work by moving from what is known or believed as a result of the processes that the constancy hypothesis supposes hold good. In other words the stimulation of the senses must on that account produce a form of perception which involves knowledge or belief. Yet neither that knowledge or belief nor the process of inference from it to the perception that finally results are such that the perceiver is in any way aware of them.

It might be objected that the process of stimulation itself need not be thought of as producing knowledge or belief but that the perceiver has to regard whatever it does produce in such a way as to make inference from it. That is to say that the perceiver has to believe something about whatever the stimulation of the senses produces as a basis for subsequent inference, but that the product of the process of stimulation does not itself have to be a belief. It sometimes happens that we infer things from what are in a genuine sense sensations. We may for example infer from the quality of a bodily sensation something about the identity of whatever is causing it, e.g. that the prick is being caused by a prickle or splinter. Such inferences, which need not be in any way unconscious, depend however on further knowledge, presumably acquired through experience, about the kinds of sensations that certain kinds of thing produce when they affect our bodies. Given this knowledge the inference to the identity or nature of the thing causing the sensation may be a genuine inference in every way. But the Helmholtzian situation is not like that (although it is perhaps worth noting that the local-sign theory about the location of sensations — a theory that Helmholtz also embraced — may involve something of this kind). Consider the Helmholtzian theory of colour contrast, which depends, as I have already noted, on the thesis that we tend to make an inference about the colour of an inset area from that of the surrounding area on the basis that we are seeing the inset area through the surrounding area. The principle of inference here must presumably be that seeing something through an area of a given colour must inevitably make the something look more like the colour of the surrounding area than it would otherwise have done. The situation would be supposedly like seeing things through coloured spectacles. But a grey patch set in the middle of a red area does not look reddish. Hence what we must infer is that its real colour is not plain grey but of a colour that is shifted more towards the opposite of red. Hence it looks greenish.

There are several points of interest in this, and a discussion of them will bring out certain general considerations about the relation between how things look and any putative intellectual processes that are involved in perception. The first thing to note is that in the story that I have given a reference to how things look appears twice. There is, first, the suggestion that something seen through another colour must inevitably look tinged with that colour. There is, second, the final report about how the thing in question actually looks, and it in fact looks different from how it would look on the first hypothesis. There is, in fact, of course the same deviation in looking at things through coloured spectacles, in that as a matter of fact things look or at least come to look more like their actual colours than one might expect at first sight. The explanation clearly lies in certain features of context. Hence the principle that is supposed to give rise to the inference is not itself founded on any confirmed law about the looks of things in these circumstances. It is based on an *a priori* belief that seeing things through a coloured medium must make them look tinged with that colour. I say that this is an *a priori* belief to indicate that it cannot in the circumstances be based on any exact empirical findings. Wearing rose-tinged spectacles may well alter our view of the world in various subtle ways but it may not make things actually look rose-coloured. Whether it *will* probably depends on a number of other factors, including how long we have been wearing the spectacles. What will undoubtedly be the case is that the wave-lengths of light reaching the eyes will be affected by wearing spectacles of this kind. If we believe that this must necessarily affect how things look, it must be because we think that there is a uniform correlation between the look of a thing and the wave-length of light which reaches our eyes from it. And such a belief is manifestly falsified by the empirical facts. Hence if we retain the belief we must do so on *a priori* grounds, and one ground of this kind is provided by the constancy hypothesis, or, to put it in other ways, by the hypothesis that how things look must be a function of the sensations that we receive, unless that look is somehow corrected by other factors.

But how can these other factors correct the look by way of *inference*, unless there is a belief connected in some way with the look? I do not mean to suggest by this that when one infers q from p, one infers q from the belief that p, although it is undoubtedly true that one comes to believe q from the belief that p, if one *has* the belief that p. The last qualification is necessary because the inference may be entirely hypothetical and need not lead to any unqualified beliefs. To infer q from p, p must be taken as a datum, i.e. as something either taken as true or assumed as true for the sake of argument.[2] Either way, it would be impossible for the inference to take place unless beliefs came into the picture in some way. The pattern of inference in the case that I am considering could only be at best: 'It looks grey. But it could look grey in these circumstances, i.e. when seen through red, only if it were in fact greenish. So it is greenish'. And then somehow it looks greenish. Here one moves from the belief about how it looks, together with the belief about its actual colour. The inference is from its look in these circumstances to its actual colour. And that is then supposed to determine its final look. If the inference is unconscious, so must presumably be the beliefs on which it is founded. Hence the first look is not one that is ever made explicit to himself by the perceiver. It is difficult to think such a view coherent.

It might be objected that the look of the thing — the final look on the previous account — is inferred not from a previous look but from the sensations that the perceiver is having. This would be like the case that I mentioned earlier, in which one infers something about the cause of the sensations from their character together with a belief that sensations of this kind are produced by things of a certain kind in turn. The inference need not of course be from the sensations to their cause; it might be from the sensations to the character of the circumstances in which they occur, without these being taken as the cause in a strict sense. Thus perhaps one might infer from the fact that one has pins and needles in one's arm that one's arm is in an awkward position. Nevertheless the form of the inference in the case before us will still be: 'I am having such and such sensations. I would have those sensations only in such and such circumstances. So the circumstances must be such and such, i.e. I must be looking at something of such and such a colour.' And then it somehow looks of that colour. In the case in question the sensations would have to be such as to be produced by light of a greyish hue in the context of light being emitted from the surroundings of a reddish hue. One would, if the inference is to be unconscious, have to believe or know this without being aware of it. If such a view is not incoherent it is at least extravagant as a possible account of what happens.

I have in all this omitted consideration of what is perhaps an even more difficult point, although it must have been evident in what I have said. This is that the inference to the real character of the object has also to determine how it finally looks. In fact in the case that I have been considering the conclusion of the inference when taken as revealing the real colour of the object is in fact false. This need not be so in all cases where unconscious inference has been invoked, e.g. distance perception where there is yet again a deviation from what might be expected on the constancy hypothesis. One sees the size of distant objects as more nearly correct than would be predicted on that hypothesis. There are no doubt cases in which how one comes to see a thing's correct size is genuinely a matter of inference. That is to say that in those cases one actually works out the size of the object from other data that are available, including perhaps an estimate of the distance or a correct view of the size of neighbouring objects. But that is not the kind of case that we are now considering, when concerned with unconscious inference; it is rather assumed that in cases in which there is no evident sign of any inference of this kind there is nevertheless inference all the same. It might well be said that even without relating this to the question of the thing's look this account of the matter is too coldly intellectual. Our ability to see things as more or less of their right size might be put down as a kind of skill. The analysis of skill is not to be carried out in terms of inferences from data. Someone who is good at estimating, say, the weight of things by looking at them need not be making any inferences from clues that he can observe, even if those clues exist. The notorious case of the chicken-sexer is crucial here; there may be clues which would tell the sexer what the sex of a chicken was if he knew about them. But there is no obvious sense in which the chicken-sexer does know them, since he has not been taught to observe such clues and go by them, and he makes no obvious inferences.

The main point on which I wish to concentrate, however, is the relation of these intellectual or quasi-intellectual processes to the look of the thing perceived; for this is the most pickwickian element perhaps in the story that I have reviewed. I said, for example, that when the perceiver has gone through the inference, unconsciously, to the conclusion 'So it is greenish', then somehow it looks greenish. Why should it? In order to get a proper view of this question we need, I suggest, to consider the different kind of factor that may affect the look of a thing and how they do so affect it.[3] One might put the Helmholtzian position that I have been considering by saying that according to it perception and how things look is a function either of the sensations that we have when our sense-organs are stimulated or of judgment, inference, or similar intellectual processes based on sensations. Since neither the inferences nor their bases are evident to the perceiver in at least the majority of cases they have to be put down as unconscious. If I am critical of this last move it is not because I have a general objection to the notion of unconscious intellectual processes of this sort. It might, for example, be necessary to appeal to unconscious inference in order to explain certain cases of mathematical insight. It is not in any case uncommon to want to say of someone that he must have inferred a certain conclusion from the data available to him without being aware that he had done so. It is in such terms that we might explain the fact that the person in question holds a certain belief when that person can give no account of how he has come to the belief. I am not saying that such an explanation must necessarily be correct; but there does not seem to be anything outrageous or logically incoherent about it. Equally there may be cases when it is reasonable to say that someone has formed a judgment on a certain matter without being aware that he has done so. Hence unconscious mental processes of this sort are not objectionable in themselves. The question at issue is whether they can be properly invoked in the case before us, in the sense that they can do the necessary explanatory work.

That there is a need to make a distinction between at least two forms of perception — the epistemic and non-epistemic — and a similar distinction between the looks or appearances of things is now relatively common-place among philosophers. (One might mention Roderick Chisholm's *Perceiving* and Fred Dretske's *Seeing and Knowing* in this connection, though I might say that an acceptance of the distinction does not commit one to agreement upon all the uses to which the distinction has been or may be put.[4]) I have spoken of the distinction as one between two forms of perception, though this may not be the best way to put the point, which can be expressed simply by saying that in some cases to see something in a certain way, or for it to look a certain way, implies a belief to that effect on the part of the perceiver, while in other cases this is not so. The distinction is perhaps most evidently applicable in the case of illusions. In those illusions which are sometimes called 'sensory', e.g. the Müller-Lyer illusion, the illusion may hold good whatever our beliefs or knowledge about the real state of things, and the illusion may not be a function of any beliefs on the part of the perceiver; it may be explicable if at all in quite different ways. (I put the matter in this rather hesitant and delicate way in order not to step on any empirical toes. I do not wish to trespass on ground reserved for empirical investigators. If the Müller-Lyer illusion is explicable in other ways, so be it; but to my knowledge the illusion is put by many psychologists in the class that I have roughly delimited as sensory. The findings of Piaget and his associates as recorded in the *Mechanisms of Perception* that the illusion is at its maximum at the youngest investigatable age are not without relevance here.) There are certainly other illusions which are the product of beliefs on the part of the perceiver. In some, if not all, of these cases the illusion can be explained by saying that the thing looks that way because for it so to look is simply, or at least in part, for the perceiver to believe it to be so, and the perceiver does have that belief. Thus, if it is true to say of a child that his father looks to him the biggest human-being in the world, this may be so because he believes this and to say that his father so looks to him is in part at least to refer to that belief. In one sense he sees his father as the biggest human-being in the world while in another sense perhaps he sees him as the same size as a lot of other human beings. I say that this may be true in some, if not all, cases where how the thing looks is produced by beliefs on the part of the perceiver, because there may be cases in which while the illusion or whatever has beliefs in its causal ancestry it itself is not to be unpacked in terms of beliefs. Such cases cannot be ruled out *a priori*.

Unfortunately, however, this qualification does not end the complexities of the situation. There are other factors apart from beliefs and purely sensory factors that may determine or help to determine how something looks. As I write this paper I can see out of the window in the distance a number of objects that look like sea-gulls. I do not know whether they are sea-gulls, and in saying that they look like sea-gulls I am not saying that I believe that they are sea-gulls; I am not even saying that I believe that they are *like* sea-gulls. But they could not look like sea-gulls if I did not know in some way what sea-gulls looked like (though it has to be admitted that someone else might say of me that the objects looked like sea-gulls to me, on the strength of a description offered by me which amounted to a description of sea-gulls, without my knowing that that was what the description amounted to). In other words the way in which we understand the world, its objects and their properties and relations may well affect how we see things and how they look to us. So much so that there is a sense in which something cannot look to us in a certain way if we have no understanding or knowledge of that way or of anything that is relevant to its being characterised in that way. In ambiguous figures like the wife and mother-in-law picture or in the duck-rabbit which, thanks to Wittgenstein, has become so popular among philosophers, there may be a variety of factors which make the picture look one way rather than the other, but it will not be seen in one way or the other if the perceiver has no idea of what a duck or rabbit looks like or perhaps if he does not know at least something of the stereotypes about wives and mother-in-laws in western society and culture. The perceiver who lacks those concepts may see the picture in some way or ways, but not in those ways in particular.

The case of ambiguous figures may also draw our attention to another factor which may play a part in the determination of how things look – attention itself. It is clear that sometimes when our attention is drawn to some feature of an object or to some part of it that object may, often suddenly, come to look to us as it has not done before. Wittgenstein devotes considerable consideration to this kind of dawning of an aspect in section xi of Part II of the *Philosophical Investigations*. Where the figure is not overtly ambiguous, as the duck-rabbit figure is, it may still be possible to see a figure like that of a simple triangle in a variety of ways, e.g. as lying down, standing up, as suspended from the apex, and so on.[5] In doing so one is of course putting the figure into something of an imaginary context, and one may be enabled to do this in a variety of ways by paying selective attention to different features of the figure. To the extent that one does this one is, as it were, making the figure ambiguous, even if it is not obviously ambiguous in itself, and the whole thing is necessarily something of an imaginative feat. It is probably for this kind of reason that Wittgenstein says that the concept of an aspect is like that of an image (op. cit. 213). This remark has sometimes been taken as casting light on the notion of an image, but I think it is meant to be taken the other way round. Thus there is also a genuine sense in which an exercise of the imagination may contribute to and partly determine the look of a thing. It was argued by Kant in his discussion of the 'schematism' and it has been argued more recently by Strawson[6] that perception always involves the imagination in some way as a link between the abstract and formal understanding involved in having a concept and the concrete experience that sense perception brings. If that is so, the consideration is relevant to the point that I was making earlier – the fact that the look of things is influenced by our understanding of the kinds of thing that they are. The kind of use of the imagination that may influence the dawning of an aspect is, I think, different[7], though it may depend on the former.

So far, I have considered factors on the side of the perceiver that may influence the way things look to him – his beliefs, his knowledge and understanding, the way in which his attention is directed, and his imaginativeness. There are of course additional factors on the side of the object or its context. A shift in what is literally a point of view may make an object look entirely different from how it has done hitherto, as may equally a change of context; it is in effect the latter which the Gestaltists emphasised so much, often to the exclusion of other factors which may in certain circumstances be just as important. What sort of account, however, can we give of the way in which these factors influence the look of a thing, when by 'look' we do not mean to imply anything about the beliefs of the perceiver with regard to the object? With this question we can go back to the kind of example that I instanced at the beginning in connection with Helmholtz. What sort of explanation should we look for in attempting to explain the facts of colour contrast? The answer may seem obvious. This is a phenomenon where context is obviously important, and we might well look for the explanation in the facts of sensory stimulation and the background physiology. That is to say that in this case the phenomenology has its basis in the interaction of neurological processes, given the way in which stimulation of the retina takes place when one coloured area is surrounded by another. It is conceivable (I say no more) that a similar reference to physiological processes will serve to account for any illusions which may be categorised as 'sensory', e.g. the Müller-Lyer illusion of it is a phenomenon of that kind.

But in the case of the Müller-Lyer illusion (and I do not know how far this applies also to other 'sensory' illusions) one can, given practice, sophistication, and determination, perhaps by directing one's attention to certain features of the figure, make the illusion disappear; the lines may be seen as of equal length. It is this kind of fact that raises in a crucial way the relation between sensory and 'intellectual' processes in perception. Helmholtz recognised this kind of point and declared categorically that where a process could be overriden by intellectual processes or by learning that process could not be a sensory one. For this reason he said that 'only qualities are sensational, whilst almost all *spatial* attributes are results of habit and experience'.[8] Such a position demands a hard and firm distinction between the sensory and intellectual in perception; that is to say that it implies that some perceptual phenomena can be classified under the heading of sensation simply, while others cannot. Those that cannot do not involve a suppression of sensations, nor strictly speaking their modification; the sensations are still there but we cease to be aware of them. To put the matter in the way in which I put it earlier in my discussion, we somehow become unaware of the fact that the area of grey inset in a red expanse looks grey in making an inference from this to the fact that it must be green. If that account was incoherent so it must be if put in terms of having sensations unconsciously and making inferences from them equally unconsciously.

I suggested earlier that instead of attempting to construe our perception of the world in terms of the ideas of data and inference or judgement, one might better invoke the notion of skill. Thus our perceiving things correctly might be thought of as the exercise or perhaps the results of the exercise of skills on our part, skills which may of course sometimes break down or be mis-applied. What I have said just now about practice, sophistication and determination would be in line with this suggestion and the direction of one's attention to certain features of the objects of perception might then be thought of as devices used in the exercise of skill. What, however, is the skill a skill at? It will not do to say that the skill is one in arriving at correct judgements about the character of things. When one is successful in seeing the lines in the Müller-Lyer illusion as of equal length one has not simply come to a correct judgement about the lines; one's *judgement* may well have been correct already. One has come to *see* them correctly; they *look* as they really are, and that look is not to be analysed simply in terms of belief. Analogously, if it is true that we have to learn to see and become good at seeing distant objects as of more or less the size and shape that they are, is it the case also that we thereby become good at estimating or judging their size and shape? There are of course occasions when we do estimate and judge these things, and there may well be devices that we can use to this end. But in such cases when we have reached a correct estimate of a thing's size we may still not be able to *see* it that way. What else is involved in *that*? Isn't it *that* that we become good at and which we learn to do as part of what is involved in learning about the world. What is it?

This seems to me an immensely difficult question, one that is this, I suspect, just because the experience is one of those, which, as Wittgenstein said, is so ordinary that we really do not find it puzzling enough. I have already referred to Strawson's discussion of the link provided by the imagination between formal understanding and experience in the phenomenon of seeing something as such and such. But whereas it is possible to understand how the imagination plays a role in enabling us to see, say, a figure in ways that we might justly call imaginative it is less clear what work a reference to the imagination plays in gaining an understanding of ordinary 'seeing-as'. It might be better to look at a very simple case of the phenomenon which is my present concern. In concluding my paper, I wish to do just this and it seems to me best to consider a case of touch, if only for the reason that touch being (a) a contact sense and (b) fairly simple in structure it does not bring in the complications of the distance senses at least.

In putting one's hand on something, and perhaps moving it over the object to some extent, one may get a variety of sensations or feelings in one's finger-tips. If one attends to those sensations under the circumstances noted one may come to find out something of their nature — whether, for example, they are closest to a tickle, or more akin to an irritation. (Unfortunately our phenomenological vocabulary is fairly restricted here, perhaps because we have little interest in this kind of operation in the ordinary way, and perhaps because there is not all that much to find out.) What happens when we come to realise or recognise that what we are feeling is, say, a fur fabric, what is it to feel it as furry, as distinct from merely having the sensations in our finger tips that I have already mentioned? What I shall offer in reply to these questions will be at the level of description only, but it may be as well to draw attention to some simple facts. First, to feel the object as furry one must have one's attention drawn to it rather than to the sensations. Do the sensations cease to exist when that happens? It seems to me difficult to believe that they do, although they do not exist as such *for* the perceiver, since they are not within his attention. (I am aware that this raises large issues, but I must leave the matter there.) Second, one must know what it is for an object to be furry, and this I take it involves knowing also in some part what it is for an object to feel furry; one might say that it involves knowing what it is for an object to be furry under the aspect 'feeling', and that one must have come to know about furriness in that way as well as in others. To know this one must have learnt what furry objects feel like. Third, one must be aware of the object and that it feels of this kind. In this the knowledge that I have mentioned under my second heading is somehow made explicit in application to the particular object to which one is attending. But the awareness is not detached from the medium by which it is mediated. It is mediated by the fingers and through the fact that the fingers are sensitive – if you like through the fact that one is having the sensations that one is having. What all this amounts to is simply the point that the form of awareness in question must be tactual, though I doubt whether this means any more than what I have already said. But there need be no beliefs about the real nature of the object or about whether it is actually as it feels. On the other hand if one does believe that it is a certain kind of object this may itself influence how it feels, perhaps because one's attention will then be directed in a way that it would not otherwise have been. For this to be possible the experience must be complex enough to admit of differences of emphasis or perspective. Hence, when I said that one must have learnt what it is for objects to feel furry, there must be presupposed in this at least the possibility of descriptions of other kinds being applied to the object in applying this one. This is a point of some importance.

I suspect that there is much else to be said here, and that what I have actually said might not add up to all that much. Some morals may however be drawn. First, whatever role sensations play in all this it is not as a basis for judgment or inference to any of the properties of the object. What might be said is that they give a kind of character to the form of awareness of the object even when one's attention is not directed to them. If one runs one's fingers over an unevenly textured object there may indeed be a kind of vacillation of attention depending on whether it is features of the object or aspects of the sensations in one's fingers that loom the larger. But when it is the features of the object that grip our awareness and attention that awareness is still coloured, as it were, by the sensations that may so easily come before our attention. Second, feeling something as such and such is in an important sense an application and manifestation of knowledge; for to feel it in this way presupposes knowledge of what it is for something to be such and such under the aspect of feeling. Yet, in order to feel it in this way it is not enough just to have that knowledge, since there are cases in which, while we know very well what the object is or what it is like, we are unable to feel it as such. There may be something lacking on the sensory side, or perhaps a lack of fit between that sensory side and the knowledge that we have. The latter is the case where, as in a puzzle picture, we may know exactly what the face to be found there looks like but just cannot see it in the picture. This may be for various reasons, such as an attention to the wrong things, and the picture may, so to speak, suddenly click into place. Sometimes there may be nothing that we can do, no steps that we can actively take, to bring about that result. But when it comes about, it is the dawning of a phenomenon which in less particular circumstances is with us all the time whenever we see something in a certain way. In some cases, as in illusions, it is in a certain sense a misapplication of knowledge, but that does not alter the general nature of the phenomenon. And if it is an application of knowledge it is not one that necessarily issues in knowledge or belief itself. That depends on whether the sensory or the epistemic side looms larger. Given this, it would be wrong to say that perception is always a matter of getting information or that what we come to be able to do when we come to be able to see something in a certain way is to be able to get information about the world. The story about perception which, à la Gibson, links it with information-getting, is at best only one side of the whole story.

Yet, third, feeling, seeing or in general perceiving something in a certain way may, nevertheless, be a source of knowledge about the world, to the extent that in so perceiving we are concentrating on or attending to the way things are or may be rather than their more aesthetic aspects on how they feel or look, in the senses of these words that I have been emphasising. It is natural that one who is concerned with the part that perception plays in biological functioning in general should emphasise that aspect of the matter, and that epistemologists should do likewise for their own reasons, but it should not be taken for granted that that is all that there is to it. At the same time, however, while the aesthetic attitude is always possible, in theory at least, and sometimes looms largest in practice, it is always theoretically possible for it to be dropped in favour of what one might call the epistemic attitude. This involves at least a shift in attention in the way that I have indicated is possible in the case of some illusions like the Müller-Lyer. And just as a belief about an object may cause it to feel or look in a certain way, so may the way a thing feels or looks cause us to believe certain things about it; there is however no necessary connection either way. Thus to the extent that we do make judgments or inferences in perceptual contexts these have no necessary implications for how things look, feel, etc., nor vice versa. (I have in the foregoing deserted the case of feeling *simpliciter*; how far considerations about feeling apply to the other senses will depend on whether there are crucial differences between those other senses and touch. There *are* differences, it is clear, but I suspect that it is a case of *mutatis mutandis*.)

Finally, a general remark about what I have been concerned with. Whatever one thinks of Helmholtz's theory of perceptual phenomena, it is clear that he worked with a very simple dichotomy of sensation versus intellectual processes. And the same applies to many other workers in this field, if with varying emphases. One thing that I have been anxious to indicate is that this is *too* simple a dichotomy. Both intellectual processes and sensations play a role in perception, but so do other things. Moreover it is sometimes the case that seeing something in a certain way is simply a matter of its looking this to us. Whatever else is to be said about this 'looking' it cannot be simply broken down or analysed into sensations, or intellectual processes like judgment, inference or belief, nor into any ordinary combination of these. If one wants to understand perception that is a point that should be faced.

NOTES

[1] Helmholtz stated this as an explicit principle. See the discussion in William James (1950), pp. 17ff, 218, 242ff.
[2] See D. G. Brown (1955).

[3] Jerry Fodor has put to me the point that since I, as will appear, allow that beliefs may have a causal role in the production of looks, there is no logical objection to invoking causality in the present case. I agree. But in that case there is no need for all the complexities of the story about unconscious inference to explain, via causal factors, the greenish look of the thing.
[4] See also the papers by J. W. Roxbee-Cox and F. Sibley in Sibley (1971).
[5] Wittgenstein (1958), p. 200.
[6] See his contribution to Foster and Swanson (1970).
[7] See Wilkerson (1973).
[8] As noted in James (1950), Vol. II, p. 218. James goes on to point out justifiably that Helmholtz was scarcely consistent over qualities and this must be evident from his treatment of colour contrast.

TO KNOW YOUR OWN MIND

All available information points to the conclusion that introspection does not exist; knowledge of mind is inferential, not phenomenal. The implications of this state of affairs however seems not to be recognized; psychologists and philosophers, many of them, still talk as if they had first-hand information about mental processes that gives them a sure basis for evaluating theory. Any such discussion today is incompetent. Mind and its activity are as theoretical as the atom, and ideas about their nature must be justified as theoretical ideas are in other fields.

I use the term introspection here in its proper sense of direct self-knowledge by the mind, the immediate awareness of its own content or activity. There is also a looser use with which I am not concerned, which refers simply to the fact that we do have a certain amount of information about what is going on in our minds at any given moment. The point made here is that the knowledge, such as it is (very limited, mostly), is the result of one or more steps of inference.

The authoritative modern critique of introspective data is by Humphrey (1951). His close analysis of the reported introspections of Titchener's group at Cornell shows that they described not sensory content, as they supposed, but the external objects and events that they perceived or imaged. No one has shown error in this analysis or refuted Humphrey's conclusion that

> we perceive objects directly, not through the intermediary of 'presentations', 'ideas', or 'sensations'. Similarly, we imagine objects directly, not through the intermediary of images, though images are present as an important part of the whole activity. (p. 129)

What one is aware of in perception is not a percept but the object that is perceived; what is given in imagination is an illusory external object, not an internal mental representation called an image. This latter, and the percept, are inferred. They undoubtedly exist, as atoms do likewise.

C. S. Peirce came to the same conclusion over a century ago (Buchler, 1940, pp. 230, 308; Goudge, 1950, p. 232; Gallie, 1952, p. 80). It is hardly necessary to say that Peirce, on this point, has not been refuted either. He has been disregarded instead. It is commonly said that Peirce was ahead of his time, with the apparent implication that we have caught up with him now, but this is a point on which he still is ahead. The point is one of fundamental philosophic significance, affecting any discussion of the nature of mind. If it is not accepted, perhaps because of its connection with other views of Peirce such as his theory of signs, there is still Humphrey's independent analysis as confirmation. It seems therefore that it is not technically competent to go on taking introspective awareness for granted, without first showing how Peirce and Humphrey, both of them, were wrong in denying its existence.

1. HOW THE INFERENCE IS MADE

Having concluded that such self-knowledge is inferential, one has some obligation to ask what the basis of inference is. This was for me a difficult task and I was concerned with it for more than a year — and then discovered that Peirce had found the answer and stated it simply: "all knowledge of the internal world is derived by hypothetical reasoning from our knowledge of external facts" (Buchler, 1940, p. 230). After I had come to this conclusion myself I realized what he was saying, but not before. I report this personal observation for its significance in showing how Peirce's idea could be for so long disregarded. If one in full agreement with the implications of the idea could fail to see what Peirce was saying, in plain English, it is not surprising that others who are less in sympathy with those implications might pass the conclusion by.

Also, Peirce did not deal with the question at length but sounds rather like a man who, having pointed out the obvious, sees no need to elaborate. Goudge (1950, p. 232–3) has summarized the argument as applied to sensation, emotion, and volition — all events that have been regarded as inescapably private. A sensation of redness is known only in the perception of red objects, clearly external to the observer; anger is something known secondarily, the primary perception being that of undesirable behavior on the part of another; and volition — this I find harder to follow — is known in the concentration of attention or abstraction, the two latter being evident in the changing and selective awareness one has of the external world.

The argument needs elaboration, and I must approach it from another direction. Let us start by recognizing that the inference to mind and mental activity concerns something happening inside oneself. The idea of the physical self, objectively known, is therefore fundamental. That idea, and the idea of the other, have been shown to have a common element (Hebb,

1960). The evidence is from adult behavior but it becomes understandable when one realizes that, for the child especially, others are perceived as integral wholes but the self can only be perceived piecemeal; even before a mirror, perspective on oneself is difficult. Inevitably then the baby must form his ideas of a *person* on the basis of other persons, objectively known, and then presumably with difficulty form an idea of himself on this model. The result is an objectively known physical self, a combination of fragmentary perceptions of his own body and the perception of others as wholes. Some perceptions of one's body are of course private — one's aches and pains and muscle tensions are not experienced by others — but they are still objective and of sensory origin: perceptions of things happening in parts of that objectively known body.

A further step comes with the idea of a *mind* as something inside the body that perceives and knows, of which the primary evidence is the peculiar relation that is evident between the posture of the body and events elsewhere. The simplest and most direct example is the relation between the position of the eye-lids and a large part of one's surroundings; at the moment when the eyes close, a segment of existence ceases, in color and shape and size. The eyes, evidently, are a sort of channel from the outside world to something within. Similarly, the ears are a channel for a different kind of information, and so with skin (for hardness, roughness, warmth), lining of the mouth (taste) and the nose (smell). The existence of that something within, mind or soul or demon, is thus inferred from the relation between certain objective phenomena.

The idea of *sensation* is the implied idea of the reception by the mind of that incoming information: an event known by inference, a highly theoretical idea and not an observable.

Of special interest, and closely related to the idea of sensation, is the image; and the first thing to note is a confusion in current discussion that derives from the fact that the term, image, is applied to two quite different entities. One is the imagined object or event that is (apparently) seen or heard or felt; the other is the theoretical activity of mind that is the basis of the experience. When I have a visual after-image, the result of staring at a bright light for a few seconds, I see before me a spot on the wall. Again there is a correlation with a phenomenon of the body, for the spot moves exactly as my eyes move, unlike other spots; the observed relation, also, between the prior exposure of the eye to the bright light and the occurrence of the spot, as well as its transience, permits the inference that what I am really dealing with is something going on inside me, an aberration of a (theoretically known) sensory process. I may refer to this internal event as a mental image, but I do not see it. In that sense of the term I *have* an image; what I *see* is the illusory spot on the wall.

My present deafness provides another example. The deaf man, especially at first, is tempted to conclude that everybody mumbles these days instead of speaking clearly (as they used to do) but it is likely to appear more parsimonious, on further consideration, to suppose instead that he himself has changed; and when he finds that he cannot hear other things that he used to hear, things that others can still hear, the latter conclusion becomes inescapable: changed properties of the objective world, as perceived, require the inference that it is his own sensory processes that have changed.

I could go on: with illusions, which are detected from conflicts of evidence from the external world, and emotions, which are recognized in oneself in the first place as they are in others, namely, by a temporary deviation from an established baseline of behavior (including changes of heart rate etc.). Secondarily, one may learn from associated signs to make the same diagnosis (Hebb, 1946; the onset of stage-fright I now recognize from a disturbed appetite and a peculiar restlessness some time before having to appear on the platform). But the examples given should be sufficient.

2. WIDER IMPLICATIONS

It is not possible logically to prove that there is no introspection – the null hypothesis cannot be proved – but Peirce and Humphrey (and Gilbert Ryle and E. G. Boring) have given us strong reason to doubt its existence, and that view is reinforced when it can be shown how inference may provide what knowledge we do have of our own thought processes. At the very least, an immediate awareness by the mind of its own activity becomes improbable and cannot be taken for granted as a basis of psychological and philosophic investigation. Let us consider some of the implications of this situation.

One concerns current discussions of what philosophers know as the identity hypothesis, the idea that mind is a part or property of the activity of the brain or part of the brain (i.e., the cerebrum). Argument against the hypothesis characteristically assumes an empirical knowledge – empirical as distinct from theoretical knowledge – of mental states and processes and proceeds by asserting that they are unlike what the identity hypothesis would require. A psychologist for example may say that a visual percept, the consciousness of some object in the visual field, cannot consist of a volley of impulses for the consciousness is not the aggregation that would imply but something forming a more smoothly integrated whole; or it may be argued by a philosopher that mind cannot be an activity of brain because mind is not localized and a brain activity obviously is. I do not speak here of the soundness or unsoundness of such theoretical ideas but only make the point that they *are* theoretical, and that the identity hypothesis is not refuted by showing that another theory is not in agreement with it.

An obvious implication, if immediate self-awareness does not exist, is an undermining of the basis of philosophic idealism. The argument here is that all we can know surely is in our own minds, so that other (external) existence becomes uncertain or known only with difficulty. In this case *knowing* becomes a relation between some part of the mind and its sense-data and not, as common sense would have it, between an organism and its environment. Idealism can certainly be maintained as a theoretical view (provided it is carried out consistently, which as far as I can make out is not done, the solipsist for example forgetting that his own body and sense-organs must be counted among the nonexistent objects of the physical world), but cannot be regarded as an inescapable consequence of the 'fact' that what we really know is our sensations. Distinctions such as that of appearance versus reality, or of the phenomenal versus the physical, as usually made, seem to be diluted forms of the same thing. It is true that the physicist's reality is theoretical and inferential and so different from the phenomenal, but the theory is based solidly on direct observation of a physical world whose reality is not brought in question — that is, on galvanometer readings, photographs of cloud chambers, and the like. The sophisticated theory that results cannot raise doubts about the reliability of these phenomena, for in doing so it would invalidate itself.

On another point, the long-standing doubts about the legitimacy of Helmholtz's 'unconscious inference' go up in smoke. There is no other kind of inference, for in this terminology 'conscious inferences' are in the class of mental processes that are open to self-inspection — of which there are none. The dichotomy of conscious and unconscious in this sense is a misconception that goes back to Herbart. It is misconception for two reasons. In the first place it assumes that one part of the mind, the conscious part, is observable, and we have seen what reason there is to deny that possibility. In these terms, there is an unconscious but not a conscious mind. This language however incorporates a fundamental confusion. It is one thing to say that I am conscious, and mean that my mind is active and responsive to my environment; it is very different to say that an idea is conscious, and mean by that that *I* am conscious of *it*. The true meaning of the term conscious is a reference to a state of the organism, as distinct from the state of a man in coma, under anesthesia, or in deep slow-wave sleep. If we accepted Herbart's (and Freud's and Jung's) terminology, we would now find ourselves in the position of having to say that man's consciousness is the result of an activity of his unconscious mind.

So what Helmholtz's 'unconscious inference' means is simply that the subject cannot report how the inference was made. This is the normal state of affairs. We have seen that one may know something of what goes on in one's mind, but the complexity of brain function is such that far more must be going on than one can know about. *All* inference is unconscious in the Herbartian sense, and there is strong independent support for that conclusion in the Würzburg studies of the imageless thought where, for example, it was shown that even elementary arithmetical operations involve an unreportable mental activity.

I have already referred to the topic of imagery, and the fallacy of mistaking the properties of the imagined object for the properties of the imaginal process itself. The nature of the process is a matter of theory, and the only evident basis for theory in this case is in the physiology of perception. In a physiological context it is possible to see the problem of imagery, and of representative processes generally, in a perhaps clearer light. It becomes evident that there may be degrees and degrees of abstraction, suggesting that we should abandon the crude categorization of representative processes as either concrete or abstract, the concrete usually being treated as some kind of image, the abstract usually as verbalization. Hubel and Wiesel (1968) provide direct physiological evidence of three degrees of abstraction in visual sensory processes (the activity of 'simple', 'complex' and 'hypercomplex' cells in visual cortex), and there are indications of a still higher level of abstraction. The cell-assemblies that I have proposed as the basis of representational activity must have a similar hierarchical order. A first-order assembly is representative of a narrow range of sensory stimulations: representative not because it resembles the stimuli but because it has the same physiological effect in the absence of the stimulation. A second-order assembly is one that represents the first-order assemblies that organized it, because it has more or less the same excitatory effect as they have on other thought processes and motor response; and so on. Now this suggests in turn that the importance of imagery is not that it carries the burden of thought after all, Binet (1903) reported an antagonism between image and thought — but that it is a *reportable* part or accompaniment of thought, consisting of first and second-order assemblies that can function as the actual process of perception does and initiate speech. And, from the point of view of this paper, it has the very important property of providing one with a clue to what is going on within one's own head: its effect is to make one see something or hear or feel something whose existence other evidence does not confirm; the conflict of evidence informs one that the something is illusory, but also provides information about what one's present thought is concerned with.

McGill University

THOMAS NATSOULAS

THE SUBJECTIVE, EXPERIENTIAL ELEMENT IN PERCEPTION

ABSTRACT. The problem of sentience and how we are to understand it is addressed. It is argued inter alia that sentience can find its place in psychological theory only in relation to perception and other modes of awareness. Basic issues examined, therefore, pertain to the subjective, experiential element in perception, to the qualitative contents of perceptual and imaginal awarenesses. These issues are basic to the theory of the percept (and image) and have to do with where and how to locate the experiential element in a materialist world view. Subjective knowledge of qualitative contents is no less structural than knowledge by description; any advantage that exists is in respect to a mode of knowing rather than in what can be known about them. The subject 'participates' in what he comes to know, but such participation does not make for knowledge unique to him. Nor does what he knows by this route count against qualitative contents belonging to brain processes, except perhaps for the problematic property of 'grain'. The discussion proceeds via review and evaluation of a variety of relevant views and arguments.

The current, concerted effort to comprehend how man processes information finds itself drawn back again to what Quinton (1970) called 'the subjective, experiential element in perception.' Hebb (1968) and Pribram (1971) attempted to give neurophysiological substance to that element in visual perception and imagery, Sperry (1969, 1970) proposed a "modified concept of consciousness" according to which the familiar colors, shapes, smells, and sounds we ordinarily take to be in the immediate environment are in fact dynamic pattern properties of certain brain processes. Recently, Shallice (1972) noted an increasing reliance by experimenters on reports of immediate experience; as a complex instance, he cited the task of adjusting a click to coincide with the termination of an image (Sperling, 1967). And in his "critique of mental imagery," Pylyshyn (1973) found it necessary, in the light of developments in cognitive psychology, to question the explanatory function of concepts of imagery, appearance, and in general the experiential element. The present discussion focuses instead on this problematic element as requiring explanation. It is that aspect of perception and imagery that is especially "program-resistant" (in Gunderson's, 1968, phrase): "For the having of pains, emotions, after-images, etc., are all

John M. Nicholas (ed.), Images, Perception, and Knowledge, 221–250.
First published in: Psychological Bulletin 81 (1974), 611–631.
Copyright © 1974 by the American Psychological Association.
Reprinted by permission

examples of non-problem-solving-non-behavior. They are not potentially well-defined tasks which hence may be programmable, for they are not tasks at all (Gunderson, 1968, p. 115; cf. Lawden, 1972, pp. 117–118)."

Perception is of course a cognitive matter. Following common usage, Hebb (1968) distinguished (a) "perceiving, ... or the process of arriving at a perception" (b) "a percept (or perception), the end product, the brain process that *is* the cognition or awareness of the object perceived (p. 468)." A percept is an achievement (as suggested by "end product") and a brain process, which as such has a temporal development and internal organization. This process was said *to be* a cognition or awareness; accordingly, the percept's being a brain process does not rule out its having a content. A subject who sees something acquires information about it. But more accurately, he acquires or tends to acquire beliefs about it; he already may have a belief that contradicts his immediate perceptual inclination to believe that something is the case (cf. Armstrong, 1968). Though there be reason on occasion to doubt and even reject its verdict, perception is the cognitive end product of those processes that keep us in informational contact with the world, including our own bodies. But perception is more than the acquisition information; this is where the subjective, experiential element comes in. Not only do we take (e.g., visually) a certain object to have a particular identity, to be of a certain kind, and to possess this or that attribute, we have a (qualitative) kind of awareness of it as "there" in a special phenomenal sense that needs to be explicated. For this purpose, an experimental demonstration is next described in which the kind of "thereness" meant is notable by its absence in the instance of one perceived object for a brief period of time.

From the instant a moving object becomes completely occluded at one end of a screen until it begins to emerge at the other end, the subject acquires and maintains, under some conditions, the immediate, attentive conviction, that the object is moving behind the screen (Burke, 1952; Michotte, Thinès, and Crabbé, 1964; Reynolds, 1968). Gibson (1966) characterized this perceptual awareness by saying that the object is "almost literally *seen*" while it travels the length of the "tunnel;" elsewhere it was observed that "the object continues to be 'seen' after it is no longer projected in the optic array (Gibson, Kaplan, Reynolds, and Wheeler, 1969, p. 114)." This phenomenological result was confirmed by Reynolds (1968), who found, under optimal conditions of entry-exit duration, that subjects reported "percepts of a figure moving continuously behind the occluding screen (p. 407)." Burke's (1952) subjects not only "saw" (as he termed it) the movement in the tunnel, they drew it without distinguishing its hidden and visible phases. The hidden phase "assumed all the characteristics of true, visible movement for its entire extent (Burke, 1952, p. 121)." Accordingly, the subjects reported that the "impression" of movement did not differ from that with the screen removed, and the impression of a stop in the movement behind the screen, produced under certain stimulus conditions, was no different than the impression of such a stop without the screen in place. Also a variety of movements were reportedly perceived depending on entry-exit conditions. Clearly subjects underwent an awareness of the object while it was completely obscured.

What kind of awareness? Burke (1952) presented these stimulus arrangements to subjects "highly skilled in observation in psychology or in the natural sciences." During occlusion, all agreed, the object's presence was nonqualitative: (a) Neither color nor form of the object was seen, and (b) though they later can form an image of the movement as a line, "at the moment of perception there is nothing of this kind (Burke, 1952, p. 137)." Michotte *et al.*, (1964) quoted a psychologist-subject expressing surprise that one can have perceptual contents of such degree of abstraction. Thus, subjects do not literally see the occluded object; the character of its presence in awareness is "amodal."[1] The object is not "there" in the qualitative sense until it emerges again from the screen. Remove the screen, or induce a powerful eidetic image (Stromeyer, 1970; Stromeyer and Psotka, 1970), and the object will be qualitatively (as well as informationally) "there" during its entire passage.

In sum, awareness of an occluded object in motion differs from awareness of a visible object in motion in respect to the stimulus-informational variables that control the two categories of awareness, but also they differ in respect to *the character of the object's presence as content of awareness*. This article examines certain basic issues that pertain to the latter, experiential element — issues basic to the theory of the percept and having to do with where and how to locate that element in a materialist world view. The concepts of awareness and qualitative content are central to the discussion; the next section is devoted to a clarification of their use.

1. QUALITATIVE CONTENT

In the study of perception, the most basic questions pertain to the nature of appearances. Early in his monumental work, Koffka (1935) asked, "Why do things look as they do? (p. 76)" and he tried to explain their looking just that way. How things look (and why) needs to be distinguished from *the fact that things do look or appear*. Thus, a question prior to Koffka's is: What is it for things to look, sound, feel, to appear as they do? Reichenbach's (1938) example of things appearing provides a concrete context in which to consider this question:

> taking a walk at dusk through a lonely moor... we see before us at some distance a man in the road. He is a strange little man, wearing a caftan, and carrying a bag on his shoulder. In spite of a feeling of uneasiness we do not doubt the man's reality. Coming nearer we see that he does not walk; he stands and waves his hand. We advance farther and discover that it is not a man that we see there but a juniper bush, a branch of which is moved by the wind (p. 198).

The man, according to Reichenbach, is a subjective thing that is "coupled" with the juniper bush, an objective thing. Both kinds of things can appear, can have "immediate existence."

Talk of things appearing is elliptical. They do not simply appear; a subject is involved — "appeared to." Being appeared to results from energy affecting the subject's nervous system. And it is tied into seeing and what is seen: "both the man and the bush have immediate existence *at the moments we see them* (Reichenbach, 1938, p. 199, italics added)." The "coupling" of bush and man is such that light reflected from the bush (or interrupted by it) determines the subject's seeing the man under certain conditions. Moreover, being appeared to is not just something that happens *when* one sees the bush or visually takes it to be a man; it happens in being aware: "Immediate existence" properly implies the manner of awareness the strange little man is "there" before the subject, he *looks to be there* in the road up ahead.[2] A bush's looking in whatever way it does can be made sense of only in relation to a perceiver, in terms of his qualitative awareness of the bush, whether as bush or as man. This is evident, too, from the nonveridical aspect: How the bush looked was itself illusory; the bush did not appear as such and was taken inferentially to be a man; later the subject recalls how the bush first looked or he reproduces that look by stepping back. The appearances of an object are its qualitative perceptual effects upon us. In the case of imaginal awareness, similar qualitative effects are produced otherwise.

Imagining a mermaid, for example, consists of a sequence of awarenesses similar to what occurs in perceiving (Hebb, 1968). Awarenesses are not awarenesses unless they have content, unless they "refer" to something and "assert" something about it (cf. Natsoulas, 1970). That is to say, in imagining a mermaid one is sequentially aware of various of her parts,

SUBJECTIVE, EXPERIENTIAL ELEMENT IN PERCEPTION 225

attributes, activities, and the like, including her "mermaidhood." Hall (1961) drew a revealing analogy of perceptions to *sentences*: Perceptions comprise "a natural language of the mind." As are conventional-language sentences, perceptions are about something other than themselves. Moreover, they are propositional in making claims about the world: "Being conscious of something – as one is in perception ... is a sort of referring to it, a meaning it of an assertive type (Hall, 1959–1960, p. 80)." Just as conventional sentences can be true or false, there are veridical and illusory perceptions. The means whereby one determines whether a perception is veridical involve drawing out its implications and testing one or more of them by performing some act or waiting for something to happen. Thus, perceptions are of a sort from which other things, namely conclusions, can follow. For perceptions to participate in the inferential process as they do, they must be propositional. O'Neil (1958), too, came to the latter conclusion by another route (cf. Rozeboom, 1961, 1972, p. 44): The content of perceptions must have "propositional form" because perceptions are a kind of knowing about, what can be known are facts, and facts clearly require sentences not individual words and phrases to be expressed.

This languagelike referring and asserting – what some authors (e.g., Perkins, 1971) call "taking" (e.g., something to be to the right of something else) – is the *minimal* concept of content, to which qualitative presence or immediate existence is in addition. Perceptual and imaginal awarenesses, in contrast to abstract thoughts (cf. Hebb, 1968), can have a qualitative dimension to their contents as well. Qualitative presence is often implied by drawing attention to the introspective likeness of visual imagery and seeing (e.g., Neisser and Kerr, 1973). One also refers to "visualization" to capture the notion of qualitative content; Hebb (1972, p. 287) suggested that people who deny having visual images can perform a certain test for imagery though introspectively their awarenesses lack "visualness." Neisser (1972) used the concept of *nonpictorial imagery* for objects imagined purely informationally. Comprehension of the concept depends on the above kind of distinction between minimal (propositional) content and the qualitative presence of that about which the awareness asserts. In effect Neisser proposed the occurrence of imaginal awarenesses of an object without its being "seen," "heard," etc., in the process; the object is imagined in such relation to another object that the latter completely seals it. Neisser and Kerr (1973) asked subjects to "visualize" several such states of affairs, a cradle inside the closed beak of a pelican, a revolver tied to a rapidly rotating propeller, a daffodil concealed in Napoleon's inside pocket, etc.

Each subject produced and rated for "vividness" six imaginal contents of this type and six of two other types. Subsequently, each content was described in response to questions mentioning only one object, the concealing object for the six "nonpictorial images."

An average 4.83 contents were described as including the concealed object, a value not different statistically from that for contents with a nonconcealing relationship between objects. But the exemplars of the concealing relationship were rated as less *vivid* on a scale from "perfectly clear and as vivid as actually seeing it would be" to no image at all. Finally, they were questioned on "precisely how the concealed object had been represented" in each instance. The concealed object was reported "visible" on two thirds of the trials, an apparent failure to follow instructions. On 19% of the trials, however, the subjects "just knew" the concealed object was there. Presumably these instances correspond to true examples of "nonpictorial imagery" in that propositional and qualitative contents are dissociated in the required way. Only on these trials is there reason to believe that the two objects (e.g., pelican and cradle) differed in how the subject was aware of them: One was qualitatively present, while the other was not. The latter object was known to be there but not, as we say, experienced.

The contents of certain awarenesses may be properly characterized as qualitative; having qualitative contents, however, does not render an awareness any less or more cognitive. A subject's inability to have certain qualitative contents does not rule out for him any piece of knowledge. Such contents make for a difference in "understanding," that ought not to be confused with knowledge:

The man who has seen and heard blue jays, who remembers their look and sound and who is capable of describing some bits of blue jay behavior, has *a kind of understanding* of blue jays which a man will lack who is capable of describing blue jays' behavior but who has the capacity neither to see nor hear blue jays (Perkins, 1971, pp. 2–3, italics added).

On the other hand, both can have equivalent knowledge including how to describe the ways blue jays appear. As Perkins (1971) stated, both can master "the theory of blue-jay-looks-to-humans and blue-jay-sounds-to humans" (cf. Hayek, 1952, pp. 32ff.). There do not seem to be good grounds to contend, as did Rosenblueth (1970, p. 67), that to a congenitally blind man "I see a red object" is necessarily a meaningless statement. It is a statement he would not use, not one he would fail to make excellent sense of had he mastered the theory of perception. The doubtless difference between the two men does not consist in what they can know or what facts they can be aware of; rather, the blind man's (or deaf man's) awarenesses lack certain qualitative contents. He would not "understand" by virtue of an inability to undergo certain processes, although he can enter the same states of knowledge otherwise.

Meehl (1966) argued rather that K_1 a sighted man, must differ in knowledge he can possess from K_2, a congenitally blind recipient of a corneal transplant who has yet to have visual awarenesses. Both possess the same, utopian scientific knowledge, including the psychophysiology of vision and the psycholinguistics of color language. Therefore, Meehl located their difference in K_2's lacking conditioned connections between visual cortex and "the tokening system," the latter defined as a

cerebral system ... which is the physical ... locus of events t_r, t_g, etc., these events being the inner tokenings of raw-feel predicates "red," "green," etc. These tokening events are the immediate causal descendents of ... events in the visual system V_3 and they are the immediate causal ancestors of events in intermediate systems that give use to families of overt acts of the reporting kind (vocalizing "red," or pressing a red-colored lever) (Meehl, 1966, p. 110).

When a red light is shone into K_2's newly unbandaged eyes, he is unable to follow the semantical rule " 'Red' means *red*." Meehl (1966) held there was a difference between the two men in their "understanding" of the semantical rule; since K_2 cannot follow the rule, he does not "understand" it. Meehl preferred to use the notion of "understanding," because both subjects have *ex hypothesi* the same utopian knowledge residing in their tokening systems (T), both, therefore *know* the rule, and more generally they can "both token T's propositions, can both derive portions of T from other portions, can both engage in metatalk about object-language parts of T, etc. (Meehl, 1966, p. 155)."

It appears that K_2 is distinguished by *what he cannot do* rather than what he cannot know. Given an auditory signal to the effect the light is red, he would be able to token "red" appropriately, albeit indirectly. However, asked to predict how a group of English-speaking people will call the light, K_2 is unable to make use of a part of T, a *known* generalization about their lingusitic habits. From this inability, Meehl (1966) concluded that obeying " 'Red' means *red*" is cognitive, for there is a cognitive payoff, an increment to knowledge, when it can be done directly and the prediction drawn. There are two points in opposition: (a) possessing any ability or skill can result in a cognitive edge, yet we would be loathe to call them all cognitive; and (b) if K_2 were informed as to the light's hue, he could make the right prediction.

He knows when his people use "red." He does not know which generalization to apply at this time, something he *can* know. Normally sighted people could be similarly situated, so far as making the proper inference, simply by not being in the room when the light was flashed.

Motivating Meehl's (1966) arguments was the feeling that one cannot know "what red looks like" unless one has had qualitative contents that exemplify red. But according to Meehl (1966), "Knowledge by acquaintance" consists, causally analyzed, in having established this interbrain linkage (tokening system to visual-cortical activity; p. 159)." The conceptual framework with which Meehl analyzed the question leads to the conclusion that a blind man can know what red looks like, because such knowledge resides in the inner tokening system, access to which is possible by more than a single means (cf. Schlick, 1969, e.g., p. 213). Additional discussion of this issue depends on clarifying what one knows when one knows "what red looks like" (see below the section entitled "A Structural Realist Thesis").

2. QUALITATIVE CONTENTS IN A MATERIALIST WORLD VIEW?

The very notion has been challenged as already inconsistent: " 'Raw feels' are by very definition outside the purview of our science (Tolman, 1932, p. 215)." Discriminations and readinesses to discriminate between stimulus objects, these Tolman allowed, but there could be no room for "unique subjective suffusions of the mind." The problem was methodological, acquiring information about the subject's "raw feels." Tolman outlined a procedure, yet argued *any* procedure, however subtle or sophisticated, would fall short of "getting across" anything more than the subject's readinesses for discriminative behaviors. One could not know whether one subject's raw feels were like those of another.[3] The subject's reports were, however, admittedly communicatory; they could produce an intricate set of relationships showing facts about his qualitative contents:

He "names" this experience. He says it is "red" of such a hue, saturation, and intensity. But suppose I am stupid and push him to the limit of his communicatory powers... (*T*)o indicate to me what the words mean... he turns at last to a color chart and points out just which colors on the chart are for him like his introspected "sensation." and which are not. And he elaborates this by indicating further just where he sees "dimensional" changes in this chart and where he does not (Tolman, 1932, p. 251).

He resorts to pointing and comparing in order to evoke in the other awarenesses with the same qualitative content as his own. Using the same means another subject could force different inferences about what may have been presumed to be the same experience. We see after all that raw feels were held to "get across," even if not in the required degree ("just what his immediate qualia, *as such*, may be"), which differs from leaving them out altogether.

Feigl (1971) would also leave something out of science, but not any event, process or feature of the world — just "the 'feel' of the 'raw feels'." What science abandons is a certain perspective, the subjective ("egocentric") one. The scientific perspective does not "participate" in its objects in the way the subjective perspective does. Whereas qualitative awarenesses are not excluded, direct access to them is *as a means of knowing them*. This kind of "double-knowledge" try at resolving the problem of qualitative contents was criticized by Sperry (1970) as a mere renaming away of the problem: Now we say only a "perspective" is ignored, nothing to be missed in our developing conception of the world. Such criticism would be well founded were it held as well that the application of phenomenal concepts is "blind" a matter of responding. When we adopt the subjective perspective on our qualitative awarenesses, however, we not only respond in ways we call knowing (Skinner, 1963). We as well "participate in the very being of a physical system (Sellars, 1971a, p. 278)." In the process of "direct labeling" (e.g., of a red "raw feel") that goes on when we engage the subjective perspective, there occurs, according to Feigl (1967, p. 155), our most immediate possible confrontation with "Being." To repeat, what science abandons is a certain perspective as such, not that perspective as subject matter.

Science yields knowledge by description, which is propositional and inferential. It reflects abstract structures of the world and its constituents (cf. Weimer, 1973). Physical systems have an intrinsic nature of which perforce we remain ignorant: "Our *knowledge* about nature is very abstract. But nature is not itself abstract (Sellars, 1946, p. 38)." This version of structural realism grants the world an inner being that we cannot know. Rosenblueth (1970) explained this limitation on our knowledge as due to the nervous system's operating on the basis of a code, phrasing in its own terms the messages received from without (cf. Globus, 1973, p. 155). These phrasings have in common with the environmental sources of those messages their structures; how we resonate perceptually to an environmental layout shares with it its structure but is in terms of a neural code. Having a common structure "implies the quantitative preservation of the relations that exist between the independent constituents an event or message through a set of transformations (Rosenblueth, 1970, p. 55; cf. Schlick, 1969, p. 159)."

2.1. Subjective Perspective

Subjective access to the qualitative content of an awareness is that awareness's propensity reliably to cause other qualitative awarenesses directed intentionally upon the content:

> That is, the "facts" of perception *qua* perception are available only (?) by employing a mode of awareness which, though much more sophisticated than the first stage of awareness we try to capture, is qualitatively similar to perceiving itself (Yolton, 1963, p. 360).

The introspective awareness of qualitative content is itself qualitative, having that content or part of it as its own.[4] Many authors have pointed out that the second, reflective stage of awareness constitutes a sophisticated activity. According to Quinton (1970), for example, it has to be "more or less arduously learnt by those who are already qualified perceivers of material objects and can be undertaken only by a special redirection of attention (p. 121)." Consider a subject at the moment qualitative aware of a tree. With a change in "attitude," as a consequence perhaps of being instructed he becomes aware of the tree's having immediate existence, of its appearance to him, of the qualitative content he has been having as such. Such changes of "attitude" or "redirections of attention" are one way to learn that a mental episode is going on, by inference from its qualitative content (cf. Hebb, 1969).

Knowledge by acquaintance (e.g., via the subjective perspective on qualitative contents) is commonly held to be noninferential (e.g., Feigl, 1967, p. 36; Meehl, 1966, p. 159), which does not imply freedom from those limitations inherent to knowledge by finite, conceptual systems. A limited repertoire of concepts and beliefs is brought to bear to better or worse effect; introspective awarenesses vary, therefore, in detail and sophistication. Consider, for example, the contrast between being "aware of no more than that something is going on in my sense-field" and being able to, "with practice, give a more or less precise and definite description of it, either in terms of colour patches or ... in terms of how things appear to me now (Quinton, 1970, p. 132)." Moreover, knowledge of our qualitative contents is necessarily structural, as is knowledge of the rest of the world:

> even in their introspective description we deal in their *structural* features. Whatever genuine knowledge we can attain is *propositional*. It reflects, for example, the similarities, dissimilarities (and degrees thereof) of the immediately experienced qualities (Feigl, 1971, p. 305).

The structural character of knowledge was explained by the fact that it is propositional (Feigl, 1967, p. 143). To stand in the relationship that they do, the knowing and the known must have something in common (cf. Gibson's concept of "the *resonance* of a retino-neuro-muscular system at various levels to the information available in optic structure (1966, p. 275, italics added)." They cannot be identical, however, for then one would have the known once and then again; and having something, however many times, is not knowing it. The knowing asserts something about the known by virtue of its structure, which is propositional; it can depict facts and non-facts (nonexistent facts) about it. The immediate "qualia" are not facts except as they stand in some relation. The propositional nature of knowledge rules out nonrelational intrinsic properties as knowable beyond being recognized, "directly labeled" (Feigl, 1967), or "indicated" (Armstrong, 1972). It is a mistake to run participation and the acquisition of knowledge together. Return to the subject's qualitative awarenesses of a tree: When they occur without implicating the subjective perspective, unselfconsciously as we say, the subject is simply participating in the qualitative contents involved. In being aware of the tree and its attributes *tout court*, one has the content (lives through it). And participation continues as one's "attitude" changes, but now another cognitive process has entered the picture: One is aware of the qualitative contents as such. Again we have the operations of coming to know, that must be in terms of some code, even though the knowledge be acquired noninferentially.

In several papers Maxwell (e.g., 1970a, 1970b) drew an implicit analogy to knowing *who* a thief is (as opposed to knowing that *someone* performed a certain robbery). In the introspective instance, we come to know certain properties of brain processes, which knowledge is analogous to knowing who the thief is upon witnessing a robbery. Twinges of pain, the visual occurrence of a red patch, a feeling of warmth, etc., are "such that we know *what* they are." But the sense of "*what* they are" can be too far-reaching though it need not be: We may not know, for example, that they are properties of brain processes, just as we may not know in a fundamental sense "what" the thief is. Maxwell would agree that any knowledge we have of the thief is structural and that our ability to recognize him as the thief does not diminish that fact.

But any direct access to our own brain processes was denied by Hebb (1968, 1969), who drew a clear line between process and content, between imaginal awareness and the thereby imagined object: When we have awarenesses "in the mind's eye" it is never images (analogous to percepts) that we are aware of, never the awareness itself, whose properties being neural are not properties of the object itself. What we are aware of is the imagined content. The brain can know its own processes only by inference.

According to Shepard and Chipman (1970), however, statements a subject makes about his present imaginal content assert facts concerning the brain process (or "representation") itself.

> it is a fact of inadequately appreciated significance that, despite the practically unlimited range and diversity of possible internal representations, we can readily assess within ourselves the degree of functional relation between any two by a simple, direct judgment of subjective similarity (Shepard and Chipman, 1970, p. 2).

Though we cannot describe the brain process involved (or even the "unique subjective experience of the color orange itself"), we can report, for example, that orange is more like red than like blue. Hebb (1968, 1969) would say that such comparisons are among attributes of imagined surfaces, from which the subject, too, has to infer the similarities between underlying processes. And indeed, the subjective perspective does not reveal the fact that any awareness is a brain process (cf. Danto, 1973, p. 410; Köhler, 1938, p. 198). The subject has no advantages in learning this over other students of his brain processes. For him to acquire such knowledge (by description) he must receive communications or make observations. It does not follow, however, that one cannot be directly aware of one's own brain processes, since one need not be aware of them *as such*. One can be aware of brain processes as qualitative contents, as occurs from the subjective perspective (cf. Globus: "The neural representations . . . within a given brain are directly known as the conscious experience with which they are identical (1973, p. 155–156)").

The distinction between process and content might help save the day in the case of abstract thought, but when the content is qualitative the distinction does not provide an escape from the problem of adequately locating such contents in a materialist framework. Hebb's (1968, 1969) avowedly monistic position requires him to identify the property of *having* a certain qualitative content with a property of a brain process. As Danto (1973) concluded, "A brain state with which . . . (a) belief (or perception) is identical would have the same content (as the belief or perception; p. 406)." Raised thereby is the difficult question to which the next section is devoted, how brain processes can have qualitative contents.

Before proceeding, a brief conclusion to this section is apropos, giving reasons for holding that qualitative contents are susceptible to intersubjective study: (a) Since differences between qualitative contents do make a difference at the very least subjectively, we can study the qualitative dimension through study of the subject's own perspective (e.g., by deriving a color pyramid). Even if his contact with his qualitative contents allows for basic description only in terms of sameness, similarity, and difference, still an ability to categorize would be at play and its products potentially available intersubjectively. (b) If having, for example, a pain "raw feel" has different behavioral effects from having a red "raw feel," then, as Sellars (1971a) suggested, "this difference can, in principle be captured by a sufficiently sensitive scientific investigation, and the qualitative dimension conceived as the one that makes *just this difference* (p. 288)." So long as there are behavioral differences, a structural characterization of qualitative contents remains possible, in terms of their causes and effects (cf. Feigl, 1967, p. 18). (c) Finally if having a certain qualitative content is a property of certain brain processes, then different qualitative contents will mean different brain processes, which can be studied intersubjectively.

3. HOW CAN BRAIN PROCESSES HAVE QUALITATIVE CONTENTS?

To accept a materialist theory of the mind, one has to pretend to be anesthetized. Locke's (1971) defense against this criticism came to a final difficulty, one he was unable to meet. Brain stimulated by electrode, he reported various splashes of color that were not properties of any object he was presently seeing. No matter how often similar sights have been previously produced by objects, "the only physical thing they can be properties of is my brain. But . . . the brain state as such does not possess these colors (p. 230)." The temptation is to dissolve Locke's problem by explaining how what occurred came to occur: This time a neurologist bypassed the normal causal sequence that begins with a source of light. The visual system was affected "higher up," and there is no more problem with what Locke "saw" than with seeing splashes of pigment on canvas, since the same mechanism is at play. This line of argument leaves the problem where it was: In the usual case the redness is a property of the paint; in the special case there is nothing whose property the redness can be.

A verbal response, too, can "symbolically represent" Locke's splash of red, which it did when he reported the latter's occurrence. But images "represent" in a way that verbal responses cannot. The splash of red was the content of an awareness in such a way (qualitatively) that it could not be the content of a response (cf. Natsoulas, 1970, p. 101; Sperry, 1952). The special way in which imaginal awareness "represent," and the existence of perceptual awareness with qualitative content, both make for the same problem: From the standpoint of the materialist conception of nature "we need to be able to make good sense of some analyzed notion of a sensuous *manner* of spontaneously taking a patch of light (and so on) to be scarlet (and so on) (Perkins, 1971, p. 336)."

3.1. A Modified Concept of Consciousness

In his attempt to make good neurophysiological sense of qualitative contents, Sperry (1969, 1970) claimed that a complete brain theory is impossible "solely in terms of the biochemical and physiological data such as we are now engaged in gathering (p. 535)." What the theory needs is certain "larger circuit-system configurations." Molar, gestalt-like, these brain processes are "different from and more than their constituent neurophysiochemical events." In part the argument for molar-level processes of this kind is based on appearances, on how qualitative contents are given to us in perceptual awareness: Such "features of cerebral excitation seem primary and essential from the subjective standpoint (Sperry, 1970. p. 590)." These large brain processes are "entitative" in how they function, in their interactions with other brain processes, each of them having "an inertia, coherence, and related dynamic properties as a unit that causes it to behave and be treated in cerebral dynamics as a distinct entity (Sperry, 1970, p. 589)." It does not suffice for them to have a certain size or complexity; they need as well "a specific operational design ... for the specific conscious function involved (Sperry, 1970, p. 589)." Qualitative contents were identified with "dynamic pattern properties" of the entitative processes: some of these "entities" in the flow pattern of cerebral excitation, for example, have a unique patterning that is the pain quality (Sperry, 1965).

These proposals are controversial (cf. Bindra, 1970), mainly because Sperry included qualitative contents *as such* in a complete brain theory. Several times one is inclined to conclude from his discussion that those special dynamic pattern properties supervene somehow their material substrate, and are in effect nonmaterial. What Sperry did hold was that the properties that make brain processes awarenesses with qualitative contents are *sui generis*. As we learn more about them, we learn more specifically about colors and sounds, for example, not about how brain processes in general function. This expansion in the domain of brain properties brings in psychic properties, but they are unique *physical* properties that only certain brain processes have. While Sperry was proposing that the familiar shapes, sounds, and colors are in the brain itself, Hebb (1968, 1969) was contending that the brain has no means for being noninferentially aware of its own

SUBJECTIVE, EXPERIENTIAL ELEMENT IN PERCEPTION

processes and treated qualitative contents as stimulus-object characteristics that the nervous system detects and notes. For Sperry (1970) they were "representations of external reality," not external reality itself. For example, where musical tones seem to be, one cannot find them even with careful study; one finds only events that cause and are "represented" by what we hear. Hebb (1972) recently suggested that awareness even of an afterimage is awareness of "something outside yourself (p. 3)." But afterimages were used as well to exemplify what is subjective; they can only be observed by the person who has them. Afterimages, too, it seems, can lead us into Locke's final difficulty.

Phantom-limb pain came up in several discussions by Hebb (1966, 1968, 1969) of "subjective evidence." Sperry (1965) used such pains to make the notion plausible that perceptual qualities are in the brain. He offered an argument from hallucination: Pains and other sensations cannot be in the absent limb, where the subject locates them, and must be, therefore, in the brain. Hebb (1968) responded to this kind of argument with a causal account of the phenomena in question: Certain nerve cells are spontaneously active in the absence of stimulation from the limb itself; a perceptual mechanis.n is operating in an extraordinary way, without the typical causal antecedents: the argument from hallucination is "faulty" because what is involved is "a mechanism of response to the environment." Generally stated, the position holds that "if we assume that sensations are *in* the mind, we could consider the mind to be observing itself. But the mechanism of sensation and imagery is a mechanism of looking outward (Hebb, 1969, p. 56)." The subject issues reports on how things look, sound, feel, taste to him, about his pains, tickles, itches, etc., and about various properties of his afterimages. All these instances (and more) involve "subjective evidence." In respect to qualitative contents, an experimenter is at a disadvantage relative to the subjective perspective and his ability to observe behavior.

However, subjective evidence implies something the subject knows. If subjective evidence is evidence, it must be evidence of something, about some entities, events, features of the world. What does the subject know? When he describes, for example, how at the moment an object looks to him, what is he telling the experimenter? Where he simply imparting information about his immediate, stimulating environment, he would have no real advantage over his audience; the latter could observe that state of affairs for themselves. In the present view (and that of Sperry, 1969, from all indications) the subject is telling about his visual processes: that one or another specific awareness or sequence of awarenesses is going on, identified in terms of its qualitative content. Hebb (1968) insisted of course that such awarenesses *are*, literally and numerically, brain processes. Thus, to be pain-qualitatively aware of an injured or diseased bodily part requires that a certain kind of brain process occur, the kind that occurs only when one is in pain. This process can be aroused more or less centrally; the "appearance" of a state of a limb can be quite convincingly produced in its absence. Subjectively the pain quality is localized in the limb. But the question is where the *phychologist* ought to locate the painful state qua appearance. Wherever the subject does? Damaged tissues are neither necessary nor sufficient for pain to be experienced. The contrasting necessity of the central process requires that the pain quality be objectively located in the brain. It is a useful, adaptive illusion that we spontaneously take pain to be at the point of injury (on those occasions that we do).

3.2. Eliminative Materialism

What brought Sperry (1969) to his modified conception of consciousness was his inability to conceive a world populated by humans and devoid of qualitative contents. Sperry found the scientific image imcomplete, needful of certain unique additions that bring qualitative contents back into it. If, as Sellars (1963) averred, "in the dimension of describing and explaining the world, science is the measure of all things, of what is that it is, and of what is not that it is not (p. 173; cf. Russell, 1946, pp. 700–701)," then one must take Sperry's problem seriously or defend the claim that as we know them there are no hues and pains, for example. Rorty (1965, 1970) offered a doctrine that bears a close relation to this claim. Eliminative materialism holds it consistently conceivable that our language for qualitative contents will be replaced one day by a language of brain processes leaving "our ability to describe and predict undiminshed." A line of criticism helps exposition of the doctrine along: Whereas pains can be sharp, intense, and throbbing, brain processes cannot (Cornman, 1968). Reporting in brain-process terms, either we could not describe our qualitative contents or our use of brain language would imply what was supposed to have been eliminated:

But if I am to describe these brain processes as *I* experience them, then I must use phenomenal predicates to describe them or if I adopt a new language, the new expressions must at least express what I now express when I report and describe my "sensations" (Bernstein, 1968, p. 271).

The counterclaim is of something there to be described, "a prelinguistic giveness about, for example, pains."

According to Rorty (1970), how such contents seem to us is not independent of the conceptual framework acquired for describing them. Given a radical change in that framework, those qualitative contents themselves would change: "if we got in the habit of using neurological terms in place of 'intense,' 'sharp,' and 'throbbing,' then our experience would be of things having those neurological properties, and not anything, e.g., intense (Rorty, 1970, p. 117)." The hypothesis of nonlinguistic awareness was critically examined as well by Taylor (1972). Following Wittgenstein, he affirmed that "for us" who are not animals or infants

> knowing is inseparably bound up with being able to say, even if one can only say rather broadly and inadequately, and even if we may have in desperation to have recourse to such words as "ineffable." An experience about which nothing at all could be said (not even this) ... would be below the threshold of the level of awareness which we consider essential for knowledge ... It would have been either lived unconsciously, or else have been so peripheral that we had or could recover no hold on it (p. 154).

But what about the kind of awareness that Cornman (1968) and Bernstein (1968) referred to, the kind on whose basis one would assign brain-process terms? Being in a position to name features of qualitative content is "already to have linguistic consciousness of one's experience, according to Taylor (1972), since "naming cannot take place in isolation outside of a context of linguistic capacity (p. 154)." But naming is one thing and being in a position to name is another. Being in such position makes linguistic awareness (naming, etc.) something that may be acquired:

> we may have a concept corresponding to an equivalence class of certain sounds or visual patterns without an explicit verbal label for it (cf. Hebb, 1949, p. 133) ... for which we *could* develop a vocabulary if communicating such concepts became important (e.g., for a professional musician, painter, or wine taster) (Pylyshyn, 1973, p. 7).

One may be in a position to say something about a qualitative content, yet be unable to, for want of appropriate words. What is the source of the desperation mentioned by Taylor, the reason for recourse to "ineffable"? Pronouncing ourselves unable to describe a content, we appear still to reach for terms whereby we might communicate (cf. Rozeboom, 1972, p. 42). How are conscious search and dissatisfaction in the attempt to match to be explained, other than in terms of nonlinguistic awareness of content? What is the match to other than the content of an awareness?

Eliminative materialism amounts to a fall-back position[5] It does not claim the linguistic replacement will occur, only that its possibility is logically consistent. There is acceptance of the contention that we do commonly have awarenesses of features of qualitative contents that are not features of brain processes in that brain theory makes no reference to them: "none of the predicates appropriate to sensations are appropriate to brain processes (Rorty, 1970, p. 114)." Does it follow that such features could not be identical with features of brain processes? The next subsection presents what seems to be the most persuasive answer to this question.

3.3. A Structural Realist Thesis

Rather than bank on the elimination of our present qualitative contents, we ought to consider the question whether they could be properties of brain processes. This means the character of our knowledge of such contents needs to be examined. The case for the structural character of introspective knowledge was presented above without specifics, without attention to just what it is we know about our own qualitative contents. Greater detail was furnished by Lewis (1966, 1972) and Smart (1971), who contended that the subject is aware of his qualitative awarenesses in terms of stimulus and response conditions that produce and result from them: "The definitive characteristic of any (sort of) experience as such is its causal role, its syndrome of most typical causes and effects (Lewis, 1966, p. 17)." For example, knowing that one is having a stabbing pain would be a matter in part of knowing that a process is going on characteristically caused by being stabbed with a pin and causing one to groan. Having an ache is known as that process which is "typically caused by such things as strained muscles, bad teeth, too much beer, and so on (Smart, 1971, p. 354)." Lewis (1966) included other "experiences" among the typical causes and effects in terms of which an "experience" is defined.

According to Lewis and to Smart, one knows as well that having a stabbing pain, for example, in a certain *kind* of process where the kind can be specified using "all sorts of platitudes of common-sense psychology," such as that toothache is also a kind of pain, that people tend to groan when they are in pain, and that pains are not like itches. The latter platitude is based on being able to compare qualitative awarenesses to note their similarities and differences. On this basis, a subject can issue reports about novel qualitative awarenesses, ones not yet known by their typical causes and effects. Smart (1971) gave the following hypothetical example of such a report: "an itchy sort of ache, stabbing towards the middle and throbbing towards the edges (p. 350)." This content is known as resembling both itches and pains, which in turn are known in terms of their causal roles, etc. In addition, there are topic-neutral properties that do not contradict the possible material character of the awarenesses to which they apply (e.g., "waxes," "wanes," "gets more intense," "comes intermittently"). Consider the experience of having an itch: The latter terms could be applied to it; it could be located in a network of resemblances to other qualitative contents; and beyond its causes and effects, that is all. This thesis implies, therefore, certain basic unanalyzable contents of which itch is an example. A subject makes comparisons between them without knowing "the respects in which these likenesses consist, though of course there must be some such respects (Smart, 1971, p. 355)." As Tolman (1922) would put it, "just what his immediate qualia, *as such*, may be" is not known to the subject himself, let alone only to him. Tolman's insistence that only similarities and differences can "get across" turns out to be a reflection of what the subject himself knows about his "immediate qualia."

Consistent with this line of thought. Armstrong (1968) argued that our concept of red — considered as a typical example of the various qualia — may be "*all* blank or gap" in that "we know *nothing* about what redness is, in its own nature (p. 275)." There is a distinct difference between recognizing red's occurrence and knowing what red is in its own nature. When necessary, we specify a qualitative awareness of red by referring to its typical stimulus conditions, listing objects we commonly take to be red, and mentioning the kind of illumination that produces awarenesses of them.[6] Because we participate in such awarenesses, however, we come under an illusion of having a "through-and-through knowledge of, or acquaintance with, such qualities as redness (Armstrong, 1968, p. 276)," an illusion that needs to be explained (see below). In the same vein, Armstrong (1972) distinguished predicates that "indicate" from ones that "analyze." His example of a predicate that only indicates was "hot"; "red" would be another. Aware by touch as we are of greater or less heat, we do not know thereby in what heat consists whether that be (in the realist mode) the mean kinetic energy of a collection of molecules or (in the equally materialist subjective mode) a property of brain processes:

What it is in fact we learn later as the result of the labours of scientists, who make a theoretical identification of heat with mean kinetic velocity of molecules. They provide us with an "analyzing" predicate. (Later science may provide still further analysis.) (Armstrong, 1972, p. 173).

Our ability to compare hues, to note resemblances between them, and to relate them in yet more complex ways presents a problem: Could we make such comparisons without knowing the respects in which the likenesses consist? Positive testimony derives from the absence of shared information about these respects especially as this absence contrasts with all that is known about the hues' structural characteristics. Note how one resorts to metaphorical expressions relating to warmth and cold, hardness and softness, etc., in order to say anything more about the hue itself (cf. Hayek, 1952). The use of metaphor brings into play still more similarities and differences between qualitative contents. If we were aware of the intrinsic character of red, there should have developed descriptive terms beyond the terms for hues. This is not to say that the subjective perspective is a linguistic one, only that the categorizations it involves ought to be expressible given time for the development of a vocabulary.

Does acquaintance notably with what cannot be defined (e.g., that through-and-through acquaintance with redness) count against qualitative contents belonging to brain processes? Thornton (1972) took issue with the idea that we do not know what red is, arguing that "knowing what colors are" is one way in which sighted and blind people differ: Without sentience, the meanings of words that signify sensible qualities and sensations cannot be "fully understood"; the phenomenal character of redness is not verbally definable; such words as "red" have "an ostensive character," meaning that one cannot *tell* someone else "*what* color red is," for example. The issue is what the subject knows once he has learned the latter ostensively. According to Thornton, the subject knows how to use the word "red" correctly and apply it to new instances. There has been no quarrel here with our knowing firsthand what red is *in the sense* that we are able to recognize its occurrence, name it, compare it with other hues, and indicate its usual conditions of occurrence. Such a concept of "knowing *what* color red is" makes agreement with Maxwell's (1972) statement possible: "We *do* know what the intrinsic (first order) properties exemplified in our sense experience are; they are properties such as *redness, warmth* (as felt), *being warmer* than, etc. (p. 135)." Thus, knowing what red is in Thornton's sense and not knowing what red is in Armstrong's sense are not far apart. Our inability to define red (except ostensively) is what Armstrong (1968, 1972) was pointing to: we know that red is occurring, but not what red is. Or as Brody and Oppenheim (1966) noted, a content incapable of description must be propertyless, an "undifferentiated unity" from the subjective perspective. Insistence on already knowing full well what red is serves the plausibility of denying that brain processes could have such contents. Examination of just what is known firsthand about red provides an antidote.

That seeming through-and-through acquaintance with redness requires explanation. An illusion of "deep" knowledge is created by participation in at least two ways. (a) What one might call "demonstrative" awarenesses occur, awarenesses that are analogous to statements such as "This is red" (cf. Clark, 1973, p. 49). They imply that redness is here and now exemplified, that it is present to be indicated. And if it is "right there," how can I not know it? Moreover, there seems to be nothing hidden about it. It does not point, in the phenomenal sense, as does an object to its other side or to its interior (Gaffron, 1956). And I can dwell on it however long I wish. How can I not know it through and through? There seems no better possible access to redness, certainly not by scientific means, with all the inference that requires. (b) The ability to conjure up a qualitative awareness of red at will provides a sighted man with a kind of "understanding" of propositions about hues unavailable to a blind man. Hearing about something being red, the sighted man can "fill in" by use of visual imagery the content of his awareness of that fact (cf. Schlick, 1969, p. 164; Perkins, 1971, p. 4). This skill leads the subject to conclude that he knows *full well* what is talked about, for he can actually produce exemplars of it that are very satisfactory from the subjective perspective.

3.4. A Difference in "Grain"

There is one characteristic of qualitative contents that makes for special difficulties in ascribing them to brain processes. Awareness of the continuous character of an expanse of red, for example, is not reducible to that awareness's causal role nor to that content's resemblances to other contents. Nor does this qualitative homogeneity fall into the topic-neutral class of predicates with "waxes," "wanes," etc., since as of now it cannot be applied to brain processes. Sellars (e.g., 1965) was led, therefore, to suggest that brain theory must include predicates that are physical$_1$ in "belonging to a spatio-temporal-nomological framework of scientific explanation and yet not physical$_2$ in belonging to a set of predicates just adequate to the theoretical description of non-living matter (p. 477)." The answer to how brain processes can have qualitative contents would be that they cannot so long as brain processes are conceived as they are now (cf. Sperry, 1969, p. 535). In order to accommodate their "grain," it is claimed that certain special "constituents" of neurophysiological states must be recognized. These constituents would not be definable in physical$_2$ terms; they would be true emergents.

It is instructive to contrast Sellar's (1965) position with that of Sperry (1969, 1970), who claimed to be proposing a controversial, emergentist hypothesis. Nevertheless, there are similarities between Sperry's analysis and Feigl's (1960, 1967, 1971) mind-brain identity thesis; Feigl has consistently maintained that physical$_2$ description of brain processes would ultimately suffice. Sperry (1970) wrote that the kind of control of the cerebral flow pattern he postulated violates none of the laws of physics, since such laws include "the fact that the position in time and space of objects and their parts is as critical in the determination of causation as are the intrinsic properties (p. 587)." The nonemergentist impression is strengthened further by Sperry's (1965) statement that a *complete* description of a mental episode by reference only to "terms of the spatio-temporal arrangement of nerve impulses," with no reference to mental properties, is theoretically admissible though impracticable. The difference in the positions of Sellars and Sperry corresponds, respectively, to the difference between (a) properties that cannot be defined in terms of properties of less complex systems and (b) properties that can be so defined but are unique to certain brain processes (cf. Sellars, 1971a, p. 284). Rather than irreducible properties, Sperry proposed properties that are for practical purposes not reduced. There is a principle of reducibility implicit in Sperry's position, one with which Sellars (1962) was fully in agreement, as follows: "Every property of a system of objects consist of properties of, or relations between, its constituents (Sellars, 1962, p. 63; Sellars, 1971b, pp. 393, 406)." But this was a major reason for Sellars to introduce as yet unknown basic constituents of neurophysiological states, since it was not evident how the patterning of neuronal activities could be identified with qualitative contents (cf. Eccles, 1972, p. 755).

This same difficulty was expressed by Feigl (1971):

There does not seem any ready explanation of the difference in "grain" between the phenomenal continuity (for example, of a small color expanse, or the homogeneity of a musical tone) and the atomic structure of "corresponding" brain processes (p. 307).

Compare this with Köhler's (1938) description of the visual field as a continuum; with regard to particulate elements (particles, molecules, neurons) he stated, "Nothing in this mosaic corresponds with the continuity of the visual field (p. 216)." On the basis of conversations with Sellars, Meehl (1966) described the perplexing "grain" as "an admittedly vague cluster of properties involving continuity, qualitative homogeneity, unity or lack of discrete parts, spatio-temporal smoothness of flow, and the like (p. 167)." Sellars (1962) made use of a pink ice cube that presents itself to us as "pink through and through, as a pink continuum, all the regions of which, however small, are pink (p. 63)." This ultimate homogeneity that (being a matter of appearance) is exemplified by qualitative contents stands in contrast to

the state of a group of neurons, (which) though it has regions which are also states of groups of neurons, has ultimate regions which are *not* states of groups of neurons but rather states of single neurons. And the same is true if we move to the finer-grained level of biochemical process (Sellars, 1962, p. 73).

How can the digital events of individual nerve impulses produce through their organization, however complex, the ultimate homogeneity of qualitative contents? Unable to conceive how, Sellars (1962, 1965, 1971b) added ultimately homogeneous "sense fields" or "sensa" to neurophysiological states. These additional constituents, Aune (1967) suggested, are no more problematical than the existence of other basic items and perhaps their existence is less puzzling than the arbitrary attachment of phenomenal properties to assemblages of neurons. Of course they are more of a problem in one significant respect: Current scientific theory has not been driven to postulate them. Therefore, it is useful to examine how Hebb's (1949, 1963, 1968) neuropsychological theory handles the "grain" of qualitative contents without strange, new constituents.

Before proceeding, some attention should be given to a proposed reason for *not* taking seriously *any* property exemplified by qualitative contents as pertinent to the question whether brain processes can have such contents. Smart (1959) distinguished the afterimage from "the experience of having an afterimage," stating that "there is in a sense no such thing as an afterimage (p. 149)." A reporting subject, in this view, does not tell us about his afterimage, that it is yellow, circular, continuous, etc., but about his "experience," which is none of those things. The ascription of these properties (yellow, etc.), Nagel (1965) added, is *no more* than part of ascribing the "having of a sensation" to a subject (Sensations exist only in being "had.") Therefore, the properties of a color expanse need be of no concern since it is not proposed that brain processes are yellow, circular, continuous, etc. The experiences or awarenesses themselves (which are brain processes) were said by Locke (1971) to be "transparent and featureless ... (T)o describe the awareness rather than the orange or the voice that I am aware of, I am completely at a loss (p. 218)." Following this line of argument, however, Locke ended up with that homeless splash of red upon brain stimulation.

Agreed having a qualitative content that exemplifies red is not a matter of being red as that is ordinarily understood. But the previous paragraph's argument runs the risk of omitting color expanses from our world picture altogether, a risk from which we may be saved by that splash of red. It forces us again to look to the brain processes *whose contents splashes of red are*: These brain processes must be conceived such that their properties are consistent with those their contents exemplify. For example, an awareness of something ultimately homogeneous must have something ultimately homogeneous about it, which does *not* mean (a) that an awareness would *look* ultimately homogeneous, or (b) that an awareness would be anything other than "transparent and featureless" from the subjective perspective. Note also that an awareness of something red, too, has nothing red about it in the sense of looking red. But there is an important difference between the two properties, redness and ultimate homogeneity: Red as experienced is a simple, unanalyzed, indicated property; ultimate homogeneity can be both indicated and analyzed — defined in terms of its structure in awareness: Each subregion of the regions of a color expanse are themselves color expanses. This definition can be applied abstractly in the search for brain process. In the case of redness, the causal roles and similarities between brain processes will have to serve alone in the search for which of them have contents exemplifying red and what analyzed property of these brain processes red is.

Globus (1973) gave as one of the problems for the present kind of view that intersubjective study of the brain will find nothing "on the order of pain, red, or orgasm." Underlying this claim is the assumption that our knowledge of these contents (pain, etc.) goes beyond our ability to recognize their occurrence in the usual way. We are supposed to know them, to be able to recognize their occurrence by unpracticed and unlearned means, observation and the like. It is very much like expecting a congenitally blind man, upon first seeing a serving of his favorite soup, to recognize it entirely without resorting to smell, taste, and feel. One should expect to find in the realm of brain processes something "on the order of red," but it will not be easy, for it will not look red. A consequence is the introspective implausibility of materialism until what red is has been determined and well learned (cf. Danto, 1973).

Hebb's primary assemblies. To characterize the contents of awarenesses, psychologists (e.g., Neisser, 1972) have used the concept of vividness. A mental episode (e.g., imagining a mermaid) that includes qualitative awarenesses is held to be more "vivid" than one that does not. "A continuum from the very vivid imagery of hallucination through the less vivid memory image to the completely abstract conceptual activity that has nothing representational about it (Hebb, 1968, p. 476)" was said by Hebb to be "implied" by his analysis in terms of lower- and higher-order assemblies. This analysis was an attempt to give neurophysiological substance to the qualitative contents of images and percepts.[7] The percept and the image both were regarded as brain processes consisting of "a sequentially organized or temporal pattern." The temporal segmentation was not arbitrary, relating as it did to the associative learning process on which mature perception was supposed to be based. Perceptual organization was supposed to be a matter of how assemblies are put together. Since each assembly corresponds to an awareness or cognition, Hebb's concept or organization is not essentially different from that of Hochberg (1970), who defined it in terms of content: "what we perceive usually has internal constraints — that is, that some features of what we say we see can be predicted from other features of what we say we see (p. 99)."

But there is an uncomfortable remainder: The eidetiker, for example, was said to see each part of the imagined object with equal *clarity*; each part is *vivid*. What makes contents qualitative so that they can be properly characterized as clear and vivid? The answer seems to be the very nature of the primary assemblies: "The hypothesis ... must be that the eidetiker has first-order visual assemblies which for some reason remain more excitable (subject to activation), for a brief period following stimulation, than those of other S's (subjects; Hebb, 1968, p. 473)." Primary assemblies were said to consist of or to be controlled by simple cells in the projection cortex (Hubel and Wiesel, 1968). If somehow only primary assemblies were active, subjects would be expected to report that they saw "something" vividly and in detail while being unable to tell what it was. Results supporting this expectation were described by Haber (1971), Haber and Standing (1968), and Liss (1968). In a series of experiments, Haber's (1971) subjects commented on the

unknowability of brief inputs. If the stimulus is terminated abruptly by visual noise, but had been bright with good contrast, subjects will say they saw the stimulus clearly, but they did not have enough time to recognize it (p. 46).

In 1963, Hebb wrote that his theory makes perception "the digital activity of cell assemblies and does not provide for analogical processes within them

(p. 21)," thus leaving certain perceptual phenomena unexplained, including some that result from the presentation of figures under stabilized-image conditions.[8] One such phenomenon is related to the "grain" of qualitative contents. Pritchard, Heron, and Hebb (1960, see also Pritchard, 1961) found that a homogeneously colored square fades in a spreading pattern rather than disappearing a distinct part at a time, as occurs in the case of outline figures. A chapter by Good (1965) then appeared in which cell assemblies were conceived as constituted of subassemblies, a modification accepted by Hebb (1968) as follows:

> The assembly itself need no longer be thought of as all-or-nothing in its activity. Fading, for example, may be a function of the density of subassemblies active in a given region (Hebb, 1972, p. 241, was much more tentative about this) ... A subassembly, conceivably, might be as small as one of Lorente de Nò's closed loops consisting of only two or three neurons (p. 471).

For Good (1965) as for Hebb (1949, 1963) the activity of a cell assembly corresponded to a "single element of consciousness." A subassembly consisted of a smaller group of neurons with "tighter" organization ("greater relative interconnectivity") that can remain active when the assembly has fatigued. The continuous "grain" of solid figures would be accounted for by Hebb (1968) in terms of the numerous such component activities in a region of the brain, despite the original conception of cell assemblies as "diffuse, anatomically irregular structures that function briefly as closed systems, and do so only by virtue of the time relations in the firing of constituent cells (Hebb, 1949, p. 196)."

Sellars' (1962, 1965, 1971b) problematic property would be treated, therefore, as a matter of dense, digital neuronal activity at the subassembly level. It is implied that the "texture" of the perceptual process may be discrete, but given numerous, closely packed constituent assemblies, the ability to detect that texture finds its limit, and color expanses *seem* continuous (cf. Pepper, 1960, p. 54, where he wrote of a "threshold of neural discrimination"). The homogeneity would be epistemological and not ontological. But all it needs to be is epistemological (a matter of awareness) for the original problem to reassert itself: The qualitative content of the introspective seeming is ultimately homogeneous (cf. Capek, 1969). Suppose that introspective awareness involves one part of the brain in some sense "scanning" another (Armstrong, 1968; Feigl, 1971). The point can then be stated as Aune (1967) did: "If the theory of the brain is a particle theory, then the expanse that somehow results from the scanning is still at odds with the gappy character of the scanner (p. 255)." This sort of unsatisfactory outcome from neural theory about qualitative contents led Sellars (1962) to introduce "sensa" as constituents of neurophysiological states. However, such supplementation of the basic items of the world was quite tentatively suggested while leaving other possibilities however indefinite open: (a) "that sensa might be 'aspects' or 'dimensions' of neurophysiological process" rather than particulars and (b) that the particles of current physical theory might end up as "singularities in a 'field' or abstractions from a domain of 'pure process' (Sellars, 1971b, pp. 416–417)." The next two subsections provide some very brief exploration of these possibilities. They are included here in conclusion because they point to the future and because they are alternatives to emergents, which materialism finds uncomfortable, to say the least. The fact that they are presently inadequate and vague does not guarantee their remaining so.

The design of slow-potential microstructure. Pribram (1971) expressed dissatisfaction with Hebb's (1949) approach as it dealt with qualitative content ("the rich tapestry of awareness"), with the neurophychological problem of "what is Imaged." An explanation is needed for "the direct immediacy of an Imaged psychological present, its existential complexity upon which the holistic Gestalt argument on perception depends (Pribram, 1971, p. 110)." Rather than take Köhler's (1958) route via neuroelectric fields with unclear relation to neuronal function, Pribram proposed a mechanism based on patterns of *neural slow potentials*. Arriving at junctions, nerve impulses produce small, slow, postsynaptic potentials that may increment in time into an impulse. In the meantime, there is possible interaction between individual slow potentials, the level of interaction pertinent to qualitative awarenesses occurring between configurations of such neuronal events. It should be stressed that the anatomical basis of these configurations is the synaptic and dendritic microstructure rather than the cell as unit. The configurations or "designs" result from (a) the presence already of a design of the same kind due to spontaneous activity, (b) the fact that postsynaptic potentials always reflect an arrival pattern (do not occur in isolation), and (c) the fact that there is time because of their size for the slow potentials to build up before there is discharge. The interaction between designs, moreover, is

enhanced by inhibitory processes and the whole procedure produces effects akin to the interference patterns resulting from the interaction of simultaneously occurring wave fronts. The slow potential microstructure acts thus as cross-correlation devices to produce new figures from which the patterns of nerve impulses are initiated (Pribram, 1971, p. 105).

The resemblance to waves must be no more than a resemblance, since the slow-potential microstructure is no less a characteristic of a pattern of digital events than is a pattern of neural impulses. Pribram (1971) made a special point of distinguishing these designs from "esoteric" or "floating" fields of the kind Köhler (1958) had proposed; slow potentials are "not diffuse but sharply localized at the junction between neurons or in the dendrites (Pribram, 1971, p. 112)." He made reference as well to the "disembodied" character of earlier concepts of "field" and "wave" as applied to the nervous system.[9] Despite the sharp localization that was believed to be a real advance, Pribram was prepared to interpret interactions among those events *in terms appropriate to wave forms* whenever the "data" required it, rather than in statistical terms appropriate to collections of individual states: "there is no special need to ignore completely a wave mechanical approach to the superposition effect (neighborhood interaction) which occurs in neural aggregates (Pribram, 1971, p. 145)." Justification for this move consisted of an instrumental attitude toward theory: The microstructures are whatever they are (characteristics of aggregates of neurons no doubt) whether described in quantal or wave mechanical terms. It is a curious position to take in view of the fact that junctional designs being equivalents of percepts is itself a piece of theory. It amounts to saying that a contradictory description may be dragged in even as the theory is first proposed.

Pure processes. Köhler (1929, p. 45) argued that since our qualitative, perceptual awarenesses depend on physical events that affect our sense organs and physiological events in the brain, and since such awarenesses make it possible for us to depict the physical world, they should allow us to draw a picture as well of that part of the physical world, the brain, to which they are closely related. The method is inference from the contents of our qualitative awarenesses to the properties of the "concomitant" brain processes. Köhler viewed these contents as a kind of window on both the physical world in general and, in this special way, on the brain. Accordingly, the characteristics exemplified in our qualitative contents can reveal some of the properties of the brain; for example, since the visual field is a continuum, its brain "correlate" must be a continuum (Köhler, 1938, p. 216). Sellars (1962) went this far, suggesting the addition to neurophysiological theory of "sensa," and was willing to go further, questioning the primitiveness of particles in the true scientific picture of the world. The subjective window raised in him the suspicion that particles might be "singularities in a space-time continuum," that there is a "nonparticulate foundation" to the current particulate image. The ultimate homogeneity of color expanses might come in time to fall in the class of topic-neutral properties mentioned above (e.g., "comes intermittently"). As far as we now know, however, ultimate homogeneity does not characterize the relevant physical processes. Sellars' (1971a) reaction was not to conclude that qualitative awarenesses constitute a separate reality; it was to attempt to "enrich our concept of the physical." Or, as Stout (1931) stated: "On this view physical phenomena and the immediate content of sense-experience are continuous in existence and fundamentally akin in their general nature (p. 251)."

Sellars (1971b) used another expression to describe how the particles of physics might end up: as "abstractions from a domain of 'pure process' (p. 416)." This mention was consistent with Aune's (1967) explication of the notion of a nonparticulate foundation as "the substitution of 'events' for things (or particles) as basic particulars of the scientific frame (p. 257)." As commonly conceived, events are not fundamental since they happen to some thing. "Events," however, are *pure* events. A world accurately described as pure process would consist of entities of all sorts as systems of "events." As Maxwell (1972) wrote, following Bertrand Russell, "the brain, like all portions of matter (would) consist of a family – or families – of events, causally related in appropriate ways ... (p. 143)." But by itself this would not get rid of the "gappiness" to which Aune (1967) referred, unless pure events and systems of them are not bounded in the way things are. We readily recongize the abstraction involved in bounding a process in time. The notion of a pure event permits a comparable recognition in respect to space: Is the event occuring only here and not there? The search would be on not for a point in space where an "event" occurs, but for the variable degree of it over space. Predictably boundaries would turn out as problematic as they now are in time (cf. Quinton: "A field is a matter of continuously varying intensities and not sharply demarcated presence and absence (1973, p. 84)").

University of California, Davis

NOTES

[1] Some comparisons are of passing interest at this point but quite relevant to the later discussion. Feigl (1967, 1971) drew attention to a significant difference between the

concepts of physical science and those of "introspective-phenomenological psychology." The former were said to be amodal, that is invariant or independent of their specific "anchoring" in a particular modality of immediate experience. Gibson (1969) pointed out the same thing for certain kinds of perceptual information: "Many distinctive features of objects and events are of this kind (corners, motions, temporal patterns, and transitions). Information for them may be extracted from more than one kind of sensory experience (p. 219)." Michotte *et al.* (1964) gave their attention to instances where such information about an object is extracted despite its being out of sight part of the time.

[2] In this connection, speaking of "looks" does not intend the resemblance of man to bush (cf. Bower, 1972, p. 52): The bush's looking like a man implies that the bush looks (appears) in a certain way. For anything to look like something else, it must appear, must look in the nonresemblance sense (cf. Vesey, 1971, pp. 8–9). It is the look of the man and a specific look of the bush that resemble each other.

[3] Hanson (1960) attacked this view as follows: If all empirical procedures fall short in principle, then the claim cannot be empirical. If there can be no way of finding out that two subjects have the same "raw feel" in some respect, then claiming that they do or do not is a claim about fact (cf. Hayek, 1952, p. 31; Schlick, 1969, pp. 175–176).

[4] Cf. Rozeboom (1972, p. 72): "introspective awareness of C may be a structure $\propto (C)$ comprising C imbedded in a special concept schema $\propto (\ )$."

[5] Smart (1972) regarded Rorty's (1965, 1970) position as a "retreat" because it seemed to him "on the face of it implausible to relegate talk of our aches, pains, and the like to the realm of talk of witches and poltergeists. . . . (p. 150)."

[6] Armstrong's (1968) analysis is a realist one (cf. Schlick, 1969, pp. 161–163) in respect to hues, which are presumed to characterize objects and light, whereas the present analysis is subjectivist in that hues are considered attributes of qualitative content (cf. Lewis, 1972, pp. 257–258).

[7] In view of recent, general criticism by Pylyshyn (1973), it ought to be stressed that images were conceived by Hebb (1968) as analogous to percepts, that is, as cognitions or awarenesses, rather than as contents of which one is aware. Thus images (or better, imaginal awarenesses) have content in the sense discussed earlier.

[8] Recently Attneave (1972, pp. 301–305) marshaled a number of results and arguments pertaining to visual perception and visual imagery to conclude that qualitative awareness of objects in space required an analog model for their explanation.

[9] Eccles (1972) joined Pribram (1971) in criticism of psychologists for being "content with such concepts as potential fields," but he was equally critical of Pribram's own speculative contribution along these lines. Eccles saw "no evidence that any appreciable action is exerted by slow potential fields as such. They are too weak to be effective . . . (and at most) can be regarded as merely being a nuisance in disturbing the functioning of neuronal systems (pp. 754–755)."

WILLIAM P. ALSTON

CAN PSYCHOLOGY DO WITHOUT PRIVATE DATA?

INTRODUCTION

The behaviorist attack on the use of introspection as a source of data in psychology is an oft told tale. It was an essential part of the early behaviorist program that psychology was to be an "objective" science of the behavior of organisms, and that as a condition of that objectivity it must draw its evidence from "public" sources. Any bit of evidence adduced must be in principle equally available to any qualified observer. This has the practical effect of ruling out private data of "consciousness" — sensory qualia, mental imagery, qualities of feeling, contents of covert thinking — which by the nature of the case are directly experienceable by one and only one person. It thereby restricts the evidential basis of psychology to reports of the overt behavior of organisms, plus facts about the physical situation of that behavior.

At first glance the behaviorist revolution would seem to have been almost completely successful in this respect, at least so far as American psychology is concerned. With the exception of a few deviants, American research psychologists generally purport to restrict themselves to public data. But to what extent does practice conform to profession? Is it really true that psychologists investigating perception, thought, imagery, memory, attitudes, and affect, eschew any reliance on their subjects' introspective awareness of what is going on in their mind? A scrutiny of the research literature suggests a negative answer. Research in visual perception heavily depends on subjects' reports as to how things look to them, research on affect depends on subjects' reports as to what they are feeling, and so on. We shall use the term "mentalistic" for research in which such reports are utilized.

Confronted with the continuing existence of mentalistic research, the behaviorist may react in either of two ways. First he may simply reject all such research as violating his ban on private data. Those who take this line turn their backs on most traditional psychological research, and seek to explore the determinants of overt behavior, construed in purely physicalistic terms. In this paper I shall not be concerned with this alternative which,

John M. Nicholas (ed.), Images, Perception, and Knowledge, 251–289.
First published in: Behaviourism 1 (1973), 71–102.
Reprinted by permission of the author and the editors.

whatever its prospects for success, at least enjoys the advantage of internal coherence.

The other, and more popular alternative, involves accommodating and legitimizing mentalistic research by reinterpreting it so as to conform to behaviorist principles. The key move in this reinterpretation is clearly indicated in the following passage from Kenneth Spence.

> ... in other words the events studied by the psychologist, Watson held, should consist in observations of the overt behavior of other organisms, other persons than the observing scientist himself, and not in the observations of the scientist's own internal activities.
>
> As everyone knows, however, most behavior scientists have continued more or less to make use of this latter type of material in the form of the objectively recordable verbal reports of their subjects. Indeed, the scientist himself, in certain circumstances, may assume a dual role and serve as both subject and experimenter. In this event his own introspective report is recorded as a linguistic response and becomes a part of the objective data. To some critics of the behavioristic viewpoint, this acceptance of the verbal reports of their subjects as a part of the data has seemed to represent an abandonment of the strict behavioristic position and a return to the conception that psychology studies *experiential* events as well as overt behavior.
>
> Such a contention, it seems to me, fails to note a very important difference in the two positions. The introspectionist, it should be recalled, assumed a strict one-to-one relationship between the verbal responses of his subjects and the inner mental processes. Accordingly, he accepted these introspective reports as facts or data about the inner mental events which they represented. The behavior scientist takes a very different position. He accepts verbal response as just one more form of behavior and he proposes to use this type of data in exactly the same manner as he does other types of behavior variables.[1]

Thus the behaviorist can accept mentalistic research, *provided* the introspective reports involved are not construed as themselves *furnishing* empirical data (concerning the subject's mental states), but are rather taken as themselves *being* public data, to *be* observed and recorded by the experimenter.

In Part I of this paper I shall argue that this behaviorist interpretation radically misrepresents the place of introspective reports in psychological inquiry, including inquiry carried on by those who subscribe to the behaviorist doctrine, and that there is no practicable alternative to regarding instrospective reports as embodying empirical data on all fours with data gleaned from observational reports of overt behavior and the physical environment. If this argument is successful, we are confronted with a choice between proscribing mentalistic research and rejecting any ban on private data. In Part II I shall argue that since the behaviorist position on data lacks adequate grounds, and since private data can satisfy any legitimate methodological requirements, we may, in good conscience, choose the second alternative.

Before launching onto these arguments we need to sharpen two of the concepts in terms of which the issues are stated.

(A) First, in order to contront the problem in its full breadth, we need a wider term than introspection, or alternatively, we need to purge that term of some of its traditional connotations. In the older "structuralist psychology" introspection was thought of as a deliberate scrutiny of one's own states of consciousness. As such it was constrasted with the naive, unreflective awareness one has of one's conscious states when one is paying no special attention to them. Now in order to concentrate my attention on a particular "content" of consciousness, there must be some discriminable item there for me to focus on, and it must stay put long enough for me to scrutinize it. These conditions seem to be satisfied by sensations of various sorts — external and internal — as well as by mental images; and it was mental contents of these sorts that received the lion's share of attention at the beginning of experimental psychology. Thoughts and feelings, however, proved more elusive, not to mention volitions and attitudes; and it is fair to say that the breakdown of introspectionist psychology directly resulted from its failure to reach consensus on such problems as that of the "composition" of thoughts and feelings, through introspective "analysis."

The "introspective reports" the status of which I will be discussing, extend wider than reports of "introspection," in the classical sense, in two respects. First they range over immediate, unreflective judgments, as well as over judgments that stem from special attention to some particular conscious content. In the usual perceptual experiment, when the subject is asked what color something appears to him to have or how many objects of a certain kind he sees, he is not expected to "introspect," to single out his visual sensations for careful scrutiny, but just to make an ordinary "naive" perceptual report. Second, the reports I shall be considering include some that it would be very implausible to regard as descriptions of any distinguishable "content" of consciousness. Consider statements by a subject as to how attractive something is to him, whether he would prefer to have A or B, how strongly he favors or disfavors ROTC, or how satisfied he is with his working conditions. Even if the traditional concept of introspection is quite in order vis-a-vis sensations and images, it seems obviously mistaken to suppose that statements of the sorts just mentioned report items located "somewhere" in the field of consciousness. The supposition that a person can concentrate his attention on a preference or a satisfaction, as he can on a sensation, and describe its qualities, was a piece of mythology constructed to bolster the claims of "introspection" to be a universally adequate approach to psychology. Nevertheless, reports of such matters share important features with, and present important problems in common with, "introspective reports" more properly so-called, and hence I shall consider them under a common rubric. To avoid the traditional associations of the term "introspection" I shall employ the monstrous neologism, "First Person Immediate Psychological State Reports" (FPIPSR's). The defining characteristics of FPIPSR are

(1) In it a person reports one of his (psychological)[2] states.

(2) It is *immediate*, in the sense of not being inferential, not based on something else the person believes. By this is meant not only that the person does not *arrive* at the report by a conscious process of inference. It is also that even on reflection he would not cite any other fact he claims to know in support of this claim. If asked "How do you know that you feel tired?", he would naturally reply "I just know.", or "It's my feeling, isn't it?"

(3) Even if it could conceivably be mistaken, the report carries a very strong presumption of truth. Given accepted epistemological standards, the person is in a position to make an authoritative pronouncement on the subject. Since we have already made explicit in (2) that the reporter has no grounds or evidence for his report (in the form of other pieces of knowledge from which it could be validly inferred), it is clear that his report is taken to be warranted, not because the reporter has adequate *evidence* for it, but simply because he is reporting what he is aware of. This is a second sense in which the report is *immediate*. It is warranted but not mediately warranted (i.e., not warranted by virtue of its logical relations to other beliefs.)

(4) No one else is in a position to make a warranted immediate judgment on this same point.

To sum up, a FPIPSR is an utterance that satisfies the following four conditions.

(1) In it a person attributes a psychological state to himself.
(2) It is immediate in that it is not based on other beliefs of the person.
(3) It is commonly taken to be (immediately) warranted.
(4) Only the person in question is in a position to satisfy all the first three conditions.

It does seem that, as a matter of fact, human beings are in a position to make FPIPSR's concerning a wide variety of their contemporaneous states, including sensations, perceptions, images, thoughts, feelings, intentions, attitudes, emotions, desires, and likings. It seems clear that I can report on what I am thinking, sensing, feeling, or intending at the moment, and that I can do so without having to find out what I am thinking or feeling via the observation of something else on the basis of which I make an *inference* as to what I am thinking or feeling. Moreover, though it would, I believe, be a mistake to reckon a person infallible on such matters, his immediate judgment would seem to carry a strong presumption of truth, and it would take extremely strong contrary evidence to overthrow it. Finally no one else is in a position to pronounce on what I am thinking, feeling, sensing, or desiring at the moment, without inferring[3] this from something else: what I say, the way I act, my demeanor or look. To be sure, not all matters of psychological fact fall within the scope of FPIPSR's. We do not generally suppose that a person is in any specially privileged position to make immediate judgments concerning his possession of such personality traits as conscientiousness or domineeringness, concerning how intelligent he is, or concerning how much musical aptitude he has. Moreover my chosen group is by no means homogeneous. It would seem that the possibility of mistake is less live, and the warrants correspondingly stronger, for the "conscious contents" traditionally handled by introspectionism than for "conative dispositions" such as desires and attitudes. Nevertheless I would maintain that there is a large class of FPIPSR's, with discernible if not perfectly precise boundaries, that can be marked off by the criteria I have provided.[4]

(B) Second, let us be more explicit concerning the precise import of the issue over the status of FPIPSR's. A short excursus into the epistemological structure of science will be useful at this point. What makes empirical science *empirical* is that statements are to be accepted or rejected in terms of whether they are supported by experience, by empirical data. To be sure, many of the statements made in scientific research and theorizing and supported by being inferred from other statements. This is true of all general statements, like "An increase in drive level increases general activity," but it is also true of many claims as to particular matters of fact. For example "Rat R_1 was hungry" would typically be supported by citing other statements as grounds, e.g., "R_1 had been deprived of food for twenty-four hours." However if *all* scientific claims were supported in that fashion, there would be no basis for considering science to be *empirical*, in contradistinction to a non-empirical discipline like dogmatic theology. It would either be just an exercise in inference, or else a matter of systematizing, and deriving implications from, a body of first premises chosen on some non-empirical basis. What gives science a title to the designation "empirical" is the fact that a given statement is considered to be supported by other statements only if the latter are supported, not by other statements, but directly by experience; or, alternatively, if the statements by which these latter are supported are themselves supported directly by experience, or.... An epistemological justification is considered valid in science only if it rests ultimately on statements that can be accepted, on the say-so of an observer, as a formulation of what he has experienced, statements that record the basic informational input into the scientific system. Following logical positivist terminology, we may term these statements "protocol statements." ("Protocol" because they constitute the record or "protocol" prepared by an observer.)

It will be noted that we have demarcated the class of protocol statements as those (1) we are *justified* in accepting *as* a formulation of what someone has experienced. We could, of course, give the term a wider meaning (2) by stipulating that it range over any statement that is put forward with the (explicit or implicit) *claim* that it is so justified. In this wider sense there could be some protocol statements that are not acceptable, e.g., perceptual judgments made under the influence of hallucinogenic drugs; whereas in sense (1) a protocol statement is, by definition, warranted. A still wider sense (3) would be that of a statement such that *if* it were issued under appropriate circumstances it would thereby by acceptable as a report of experience. A statement can be a protocol statement in this sense even though no one in fact issues it as an experiential report. Thus if in fact no one looked into the lab between 2:00 and 3:00 p.m., "R_2 ate a food pellet at 2:15" might be a protocol statement in sense (3) but not senses (1) or (2). Note that we might be justified in accepting such a statement, though not as a report of observation. That could be the case here if we had a reliable instrument that recorded the time of R_2's pellet ingestions.

All of these senses determine interesting classes of statements, each of which is worthy of attention. Moreover, I am anxious to stress the difference between senses (1) and (2), because I want to emphasize the point that I do not take a putative protocol statement to be acceptable *just because* it is confidently put forward as such, or *just because* it seems to its proponent to correctly formulate what he is experiencing. The scientific community (more generally, the "epistemological community") quite properly reserves the right to reject a statement put forward as a protocol statement if the conditions are not appropriate. If a person were not in a good position to determine just exactly what happened when the cars collided, or if we have good grounds for supposing that his perception of the occurrence was influenced by his personal bias, we may well throw out his testimony. However, for the purposes of our discussion in this paper, we can restrict ourselves to using the term in sense (1).

It is by no means an uncontroversial matter just how the class of protocol statements is to be demarcated. In practice, ordinary perceptual judgments like "Rat R_1 is eating rapidly" are so treated; i.e., scientific investigators proceed as if a qualified observer, properly situated, is justified in making such a statement as a report of what he is perceiving, even if he is not able to infer it from other acceptable statements. Philosophers of a phenomenalist bent, on the other hand, maintain that a more refined analysis of the epistemological structure of empirical knowledge would reveal that such physical object perceptual statements are justified only to the extent that they could be inferred from "phenomenal" statements, each of which simply reports the content of the perceiver's experience without making any commitment as to the existence of anything outside that experience. From such a point of view the above rat-statement is not a protocol statement. Again, psychoanalysts seem, in practice, to take such statements as "At that point the patient became very hostile" to be protocol statements, a practice condemned by many methodologists. But wherever the line is drawn, our present point is that it is essential for the status of science as an empirical discipline that the line be drawn *somewhere*[5]

The area of epistemology and philosophy of science in which we have been treading is full of conflicting positions as well as profound problems. Many philosophers would not accept the claims we have just advanced, at least not without qualifications. Again, there are subtle and difficult problems as to how phrases like "formulating *what* is experienced," are to be understood. Unfortunately we will not be able to go into these matters in this paper.

Using these terms we can formulate the behaviorist's ban on private data as the thesis that FPIPSR's are not to be allowed to function as protocol statements in psychology[6]. And the issue concerning mentalistic research can be formulated as follows. "Should we regard the FPIPSR's issued by the subjects in mentalistic research as protocol statements?" Traditionally they have been so regarded, and it is quite natural to do so. It looks for all the world that when the subject in a perception experiment says what color the object appears to him to have, the experimenter accepts that statement just as a report of his experience, and uses it as a piece of evidence in terms of which his hypotheses are tested. But on the behaviorist reinterpretation the FPIPSR is not taken as a protocol statement. In fact it is not treated as a (true-or-false) statement at all[7]. It is not put into the network of scientific inference. The behaviorist, as Spence says, will use it "in exactly the same manner as he does other types of behavior variables"; that is, it is treated as an event to be *observed* and reported on, rather than as itself a report of an event. The empirical base is not the FPIPSR issued by the subject, but rather the experimenter's report of that utterance. It is their reports of the subject's reports that constitute, for the behaviorist, the basic protocol statements. We shall nor proceed to consider whether the behaviorist reinterpretation is valid.

PART I

(i) Our discussion of this question will be more concrete if it is carried on with reference to particular investigations in which FPIPSR's figure. For this purpose I choose one case from the study of perception in which the putative private data are the sorts of things traditionally regarded as objects of introspection, and one case from cognitive dissonance research, in which the data are not of this sort.

Our first example comes from an article by J. E. Hochberg, W. Triebel, and G. Seaman entitled "Color Adaptation Under Conditions of Homogeneous Visual Stimulation (Ganzfeld)."[8] Part of their aim was to test the hypothesis that "fields of colored illumination of sufficient homogeneity to be perceived as surfaceless fog will lose their color and become chromatically neutral" after prolonged inspection (p. 154). Upon prolonged exposure to homogeneous red and green visual fields almost all their subjects "experienced a stable, well-defined disappearance of the color" (p. 156) within less than six minutes, thus confirming the hypothesis.

The second example is taken from J. Brehm's "Post-decision Changes in the Desirability of Alternatives."[9] His hypotheses derive from Leon Festinger's theory of cognitive dissonance, and he formulates the first two as follows (p. 384).

(1) Choosing between two alternatives creates dissonance and a consequent pressure to reduce it. The dissonance is reduced by making the chosen alternative more desirable and the unchosen alternative less desirable after the choice than they were before it.

(2) The magnitude of the dissonance and the consequent pressure to

reduce it are greater the more closely the alternatives approach each other in desirability.

In the study through a series of subterfuges, the details of which do not concern us, subjects were induced to rate a number of different articles – a toaster, a portable radio, a stop watch, a desk lamp, etc. – on an eight-point scale ranging from (1), "definitely not at all desirable," to (8), "extremely desirable." At a later stage each subject was led to believe that she could have her choice of a pair of articles. In each case the pair was chosen so that they were either quite close together on her original rating (high dissonance condition), or rather widely separated on the original rating (low dissonance condition). At a still later stage each subject once again rated all the articles on the same desirability scale. Data relevant to the two hypotheses cited are presented in the following table.

	N	Average Initial Rating	Average[10] Rating Change
Low Dissonance Condition			
Chosen article	33	5.98	0.33
Unchosen article	33	3.54	−0.14
High Dissonance Condition			
Chosen article	27	6.19	0.20
Unchosen article	27	5.23	−0.66

These figures are then subjected to various statistical manipulations, the details of which need not concern us. The point is that the hypotheses are supported by showing that the chosen alternative gained in desirability and the rejected alternative decreased in desirability after the choice, and that this shift was, overall, more pronounced in the case where the alternatives were relatively close in desirability at the outset.

(ii) Now the first thing to note about these experiments is that as they are conceived and presented, the FPIPSR's involved have to be construed as protocol statements. Not that the authors themselves make any explicit pronouncements on this point. The point rather is that they must take their subjects' FPIPSR's as protocol statements *if* their hypotheses as stated are to be supported by the data. Brehm's first hypothesis is that after a choice between two (desirable) alternatives the chosen alternative will (tend to) *become more attractive* and the rejected alternative will (tend to) *become less attractive*. That is, the hypothesis is stated in terms of a shift in *degree of attractiveness* after the choice (not in terms of how rating behavior will be affected). If the individual data out of which the statistical constructions are built consist of such facts as that "Prior to the choice the toaster was only moderately attractive to S_1" (i.e., consist of data reported by the ratings), then the data have a clear evidential bearing on the hypothesis. For the statistical manipulations of such data reveal that there were pronounced shifts in the predicted directions. If however the individual data consist of such facts as "Prior to the choice S_1 gave the toaster a rating of 4," there is, on the basis of considerations presented, no such evidential relation. For now there is a conceptual gap between the hypothesis and the data. The hypothesis predicts a change in degree of attractiveness under certain conditions; the data embody a change in ratings under those conditions. Before we can take *these* data to support *that* hypothesis, some bridge will have to be built between facts about such linguistic responses of subjects, and facts about the attractiveness of articles to those subjects. No such considerations are presented. Brehm does nothing to indicate how data concerning rating behavior can support hypotheses about degrees of attractiveness; he seems to suppose that his data *as they stand* support his hypotheses. Hence he must be taking his data to consist of facts about how attractive a given object is to a person at a given moment, i.e., facts of the sort reported in the FPIPSR's, not facts about verbal behavior of the sort reported by the experimenter.

A precisely parallel point can be made concerning the other article. The hypothesis concerns what will happen to visual perception under certain conditions, not what experimental subjects will say under certain conditions. Thus, once more, if and only if the data have to do with how something looks (the sort of thing reported by the subjects) the data have a clear evidential bearing on the hypothesis. I take it that in this respect our examples are quite typical of mentalistic research.

Thus, if the behaviorist is going to construe mentalistic research in such a way that the experimenter's reports of the subject's reports, rather than the subject's reports function as protocol statements, he will have to do something to bring his version of the data into an evidential relation to the hypotheses. (Otherwise, on his interpretation, all mentalistic experiments will turn out to be worthless.) There are basically two ways of doing this.

He can modify the hypothesis, or he can build a bridge from (his version of) the data to the original hypothesis. Both possibilities are suggested by Spence, immediately after the passage quoted earlier. After saying "He accepts verbal response as just one more form of behavior and he proposes to use this type of data in exactly the same manner as he does other types of behavior variables," Spence continues:

Thus he attempts to discover laws relating verbal responses to environmental events of the past or present, and he seeks to find what relations they have to other types of response variables. He also makes use of them as a basis for making inferences as to certain hypothetical or theoretical constructs which he employs. In contrast, then, to the introspectionists's conception of these verbal reports as mirroring directly inner mental events, i.e., facts, the behaviorist uses them either as data in their own right to be related to other data, or as a base from which to infer theoretical constructs which presumably represent internal or covert activities of their subjects.[1][1]

The first possibility involves attempting "to discover laws relating verbal responses to environmental events of the past or present, ... and to other types of response variables." Applied to the present problem, this would involve getting the hypotheses into connection with public data by *altering the hypotheses* so that the dependent variables are overt utterances, rather than private states or experiences.[1][2] The second possibility involves using the subject's reports "as a basis for making inferences as to certain hypothetical or theoretical constructs ... which presumably represent internal or covert activities of their subjects." Here we leave the hypotheses in their original form and attempt to build some kind of inferential bridge from the occurrence of FPIPSR's to the inner states or experiences specified in the hypotheses. We shall examine these two suggestions in turn.

(iii) On the first kind of reinterpretation, the first Brehm hypothesis would be restated so as to stipulate that after a choice between alternatives both of which were *rated* as desirable one would, if he were set the task of rating the alternatives again, tend to *rate* the chosen alternative higher and the rejected one lower. And the perceptual hypothesis would become: When a subject is continually exposed visually to a chromatic Ganzfeld, then if he is asked after about three minutes how it looks, he will reply that it no longer looks colored.

Now these "behaviorist transformations" are not, of course, supposed just to be *other* hypotheses that it would also be interesting and fruitful to investigate. The claim is that they capture, in an acceptable form, the (legitimate) substance and real scientific import of the original. The idea is that by means of this transformation we will preserve everything in mentalistic research that is of value to psychology as a science, while at the same time avoiding any appeal to private data. To fully appreciate why this has seemed plausible we have to bring out into the open the operationalism that has historically been closely associated with behaviorism.[13] In the experiments under discussion, FPIPSR's constitute the experimenter's access to the mental states he says he is studying. Therefore, these overt indicators will provide our "operational definition" for the "inner mental states;" they are what the inner mental states operationally amount to. From an operationalist point of view it was overt verbal behavior that was really being used all along as the dependent variable in these experiments, a fact that has been obscured by the traditional mentalistic framework within which the research has usually been conceived. Thus the proposed "reinterpretation" will simply be an explicit recognition of the true character of the situation and will in no way diminish the value and significance of the results.

Now what would it take to make a behaviorist transformation equivalent in "legitimate scientific import" to the mentalistic original? Naturally we cannot require that they be equivalent in every respect; in that case we should have no *revision*. Without attempting to lay out *sufficient* conditions for such an equivalence, I think we can discern two necessary conditions that are quite relevant to our present concerns. Where, as in these cases, the two hypotheses differ by the substitution of one variable for another, the hypotheses will have the same "scientific import" only if (a) the two variables are extensionally equivalent and (b) fit into the surrounding body of theory in the same way. (a) is required, because if two variables are not extensionally equivalent — if they do not always occur (or fail to occur) under the same conditions and do not always give rise to the same consequences, then clearly hypotheses that spell out their functional relations to other variables cannot amount to the same thing. (b) is required because if it is lacking, the hypothesis will not play the same role in such activities as explanation and theory construction.

Having detected the operationalism lurking behind this first behaviorist move we should be prepared for the fact that, like other "operationalist definitions," these revisions fail to satisfy both the above conditions. I shall now proceed, in the case of each, to show that these expectations are borne out.

(a) As we have formulated the behaviorist transformations, the new dependent variables are *obviously* not extensionally equivalent to the originals, for they are obviously not governed by the same conditions. The conditions under which something will look achromatic to a person, S_1 are by no means identical with those under which S_1 will say that the thing in question looks colorless. (And the conditions under which an object seems desirable to S_1 are by no means identical with the conditions under which S_1 will rate the object as desirable.) For one thing, there is the obvious point that one need not communicate one's visual impressions or attitudes; most of them, in fact, go unreported. However the reformulations presented above attempt to take care of this by further enriching the hypotheses to include in the dependent variable the stipulation that the subject was requested to give a report or rating. But even so, whether any report is issued and, if so, what its content will be, is affected by variables quite other than those that can reasonably be expected to affect visual appearances or degrees of attractiveness. Whether I say anything at all about how the object looks to me depends, e.g., on whether I understand the experimenter's instructions and whether I am disposed to cooperate with the experimenter, while neither of these factors will have any significant influence on how the sphere looks to me. Again, if I do answer, *what* I say will be influenced by a variety of factors that again will presumably not influence the visual appearance itself, e.g., my verbal resources for reporting visual impressions, my inclination to be truthful or mendacious, and the care with which I choose my words. This clearly indicates that under certain conditions the manipulation of visual stimuli will give rise to a certain visual impression without giving rise to a report of that impression; hence the two hypotheses clearly do not come to the same thing.

"Be all that as it may," the behaviorist may rejoin, "the hard fact is that in the sort of research under investigation the correlations actually discovered are between antecedent conditions and verbal reports of subjects. Therefore if we were justified, on the old scheme, in thinking that the inner mental states were under the functional control of the antecendent conditions specified, the empirical results clearly show that the verbal reports are under the control of these same variables. Hence, however plausible your argument, the empirical facts show that your conclusion is mistaken."[14]

Now there can be no doubt that mentalistic research does give support to the claim that in the typical experimental situation verbal reports are under the functional control of the antecendent variables studied. But the qualification just made, "in the typical experimental situation," is crucial. What this research shows is not that verbal reports are *generally* correlated with the antecedent variables studied, but rather that they are correlated when certain special conditions are satisfied. These conditions parallel the factors I earlier spoke of as affecting the verbal reports but not the "inner mental variable." The experimenter is careful to make clear to the subject *what* he is supposed to report or *what* question he is supposed to answer. He sees to it that the subject is motivated to give him a report on the subject. If there is any reason to suspect carelessness or mendacity, this is guarded against in advance, or the data are thrown out. In other words, the Ganzfeld experiment, e.g., gives evidence for a correlation between certain physical conditions and a *report* to the effect that the sphere looks achromatic, only if the further conditions are satisfied that (1) the subject understands that he is to say how the sphere looks to him with respect to color, (2) he complies with this instruction carefully and honestly. Thus if these conditions are added to the consequent of the behaviorist transformation, it is not implausible to take it as extensionally equivalent to the original consequent.

But now notice the position into which we have worked ourselves. These additional conditions are themselves stated in terms of the notion of the subject *saying how the sphere looks to him*, i.e., in terms of the subject giving a report on his private visual experiences. That is, in the very act of trying to reformulate the hypothesis with verbal utterances as the dependent variables, we find ourselves forced to use just the concept we were trying to avoid; we find ourselves forced to think of the verbal report as a (true or false) statement concerning the subject's visual experience. Without thinking of the report in this way we cannot specify what instructions he has to be following in order for his utterance to count as a relevant datum. And without thinking of what he says as a statement that can be assessed as true or false, we cannot make sense of the condition that he be speaking honestly (i.e., saying what he believes to be true) and that he be proceeding carefully, i.e., trying his best to make an *accurate statement*. The traditional interpretation has re-entered by the back door. In order to make verbal utterance variables (even roughly) extensionally equivalent to experiential variables, the behaviorist must pay the price of hedging the former with conditions that presuppose the unwanted interpretation of those utterances as FPIPSR's. If we were able to specify in physicalistic terms the conditions under which verbal utterances could be expected to correlate with the specified antecedent conditions, we could avoid this difficulty. But so far as I know, no one has provided any such formulation, and I, for one, do not see how it could be done.

In fact the reversion to the traditional interpretation can be discerned at a more basic level. Even apart from the conditions that have to be *added* to the consequent of the hypothesis, there is the problem of how to construe the verbal-utterance variable itself; and it would seem that any workable way of doing so would again involve the notion of a (true or false) report of private data. Just what is it, in the way of verbal utterance, that we can expect, under the conditions mentioned, to issue from looking at a chromatic Ganzfeld for three minutes? The natural way to specify this is as I have done; the expectation is that the subject will *report that the sphere looks achromatic*. But that way of specifying the dependent variable again involves thinking of the utterance as a report of a private visual impression, just the interpretation we are trying to avoid. An obvious move by the behaviorist at this point would be to retreat to a lower level of data-specification. When we think of the subject's verbal behavior in such terms as "He reported that the sphere looked achromatic," we are conceptualizing it at the "illocutionary act"[15] level: we are making explicit *what* he said, the "content" of his utterance. Such a description is at a realtively high level of conceptualization because it abstracts from any specification of the particular symbolic device employed (what sentence in what language is uttered), and it is noncommittal on various linguistically accidental features of the performance — tone of voice, accent, decibel level, etc. Instead it makes explicit the meaning or import of the utterance, however realized. However it is obviously possible to report verbal behavior in such a way as to include what the illocutionary-act report excludes, and vice versa. That is, we could report our subject's utterance by making explicit what sentence or other symbolic device he employed ("He uttered the sentence, 'It doesn't look colored any longer' "), ("locutionary act" level), or by making explicit the physical character of the sounds he produced vocally ('He vocally produced a sound-sequence of the following acoustical character — '), ("phonetic act" level), while avoiding any specification of the content of what he said. Now if the verbal-bahavior variable is conceptualized at the locutionary or phonetic-act level, the behaviorist escapes the above objection. For in specifying what physical sounds the subject produced, the experimenter is not thereby committed to regarding the subject's utterance as an introspective report.[16]

But this latter alternative is itself exposed to a devastating objection, based on the indefinite variety of locutionary devices that can be used to make any given report. To report that something looks achromatic a subject might produce one or another set of noises, constituting one or another sentence. If his only language is something other than English he will not produce the above sentence, even if all conditions so far mentioned have been satisfied. Even within English there is a large number of different sentences that could be used to give a report of the sort predicted. These include: "It seems achromatic," "It looks light gray," "I don't see any color now," "The red color has disappeared," etc., etc. Let us suppose that in existing human languages there is only a finite, though undoubtedly very large, number of standard ways of reporting that a certain sphere no longer looks colored. In that case it seems that we could replace the illocutionary-act dependent variable, "The subject reports that the sphere no longer looks colored" with a very large disjunction: "The subject says 'The sphere looks gray now,' or he says 'La sphere ne semble plus rouge,' or he says..." Needless to say, such an hypothesis would be most unwieldy.[17] But that is not the worst of it. It also faces a fatal difficulty. Language is essentially an open-ended affair. Within a single language there are resources for developing an indefinite number of new ways of saying something or other (through metaphor, introduction of new terminology, shifts in the meanings of words, etc.); and of course no limit can be set on the development of new languages and new language-derivatives like codes. This means that if our hypothesis is stated in terms of a finite disjunction of locutionary-act types, it can be falsified at will simply by introducing a new way of saying that a sphere no longer looks colored. For example one can devise a special code for the purpose and in accordance with that code, make the appropriate report by uttering the sentence "The moon is blue." This utterance was not included in our disjunction, and so his saying *that*, under the specified conditions, falsifies our revised hypothesis. But the same result would confirm the original hypothesis. For the subject would be reporting, albeit in neologistic fashion, that the sphere no longer looked colored to him. And no matter how large we make the disjunction we will still be vulnerable to such moves. What this shows is that no formulation in locutionary or phonetic terms, however complex, will stand a chance of being extensionally equivalent even to an illocutionary-act dependent variable, much less to the orginal experiential variable.[18]

To sum up the discussion of this first requirement, in order to make his overt utterance variable (even approximately) extensionally equivalent to the original, the behaviorist is forced to construe the subject's report as a careful, honest report on the content of the subject's conscious experience — just the construal the reinterpretation was designed to avoid. The price of extensional equivalence is to wash out the difference for the sake of which the reinterpretation was attempted.

However, there is a complication which makes this argument less conclusive than has just been alleged. What the behaviorist specifically wants to avoid is treating the subject's report *as a protocol statement*. Might he not plead that he can construe the report as a statement by the subject about the content of his visual experience (a statement which could, if one wished, be evaluated as true or false), without taking it *as a protocol statement*, i.e., without himself (the experimenter) making use of *it* as part of the evidence for his hypothesis? After all, there are many psychological investigations into the determinants of verbal utterances, in which the investigator recognizes these utterances to be (true of false) statements, without thereby committing himself to including what is reported in these statements among his data. Suppose a psychologist is investigating the way patients perceive their relations to their analyst. In the course of this inquiry he tests the hypothesis that oral-dependent types are more likely to report that their analyst does not offer any helpful suggestions. Now in testing this hypothesis the investigator will, of course, record the frequency with which such a report occurs in various samples of patients. And he will, no doubt, recognize these reports as statements that could be evaluated as true or false. But that by no means commits him to using those statements as protocol statements. That is, he is not committed to taking the putative facts reported in these reports as empirical data for the evaluation of his hypothesis. And obviously he will not do so. Why can't the behaviorist escape our dilemma by taking a similar position here?

A behaviorist who made this move would not be contradicting himself. The only question is as to whether it would be a reasonable line to take in *this kind of case*. No doubt it is often reasonable, as in the above clinical investigation, to make use of subjects' reports without using them as protocol statements. But how about the present case? Is it reasonable, when investigating the determinants of visual sensation, to recognize that one is recording one's subjects' careful and honest attempts to give an accurate description of the character of their visual sensations, and yet to refuse to use the facts reported by those subjects as evidence? I should think not. I should think that once the behaviorist has been forced out of the extreme position of regarding "verbal response as just one more form of behavior," once he has been forced to regard these "verbal responses" as deliberate attempts on the part of the subject to give an accurate description of these visual experiences, then (unless he has some reason for doubting their prima facie reliability) there is no reasonable stopping place short of using these reports as protocol statements that furnish data for the evaluation of his hypotheses. However, I do not wish to insist on this point. Let us admit that this first argument is less than conclusive, and pass on to a more conclusive argument, based on the second necessary condition.

(b) The second requirement, let us recall, was that the behaviorist transformation should fit into the larger context of psychological research and theory in the same way as the mentalistic orginal. With respect to the first requirement, my argument was not that it could not be satisfied, but rather that it could be satisfied only at the cost of reverting, in effect, to the mentalistic interpretation. Here my argument is more simple; the revised version is simply not equivalent in this respect.

This can be seen in several ways. First, there is the point that mental states can be manifested in ways other than by explicit verbal reports. Consider the Brehm study. We have ready at hand a variety of devices for determining how attractive some object seems to some person, other than getting the person to deliberately indicate this to us in some way. Pupil dilation has been discovered to vary lawfully with positive interest in an object looked at. And in principle we could observe the subjects' spontaneous behavior in the presence of the objects in question in order to get estimates of degrees of attractiveness. Suppose that we tested the Brehm hypothesis, using pupil size rather than rating scales as a measure of degree of attractiveness. In that case we would be testing just the same hypothesis (as formulated by Brehm); but obviously it could not be replaced here by a hypothesis that takes verbal reports as a dependent variable. Just because degree of attractiveness can be gotten at through manifestations other than testimony, an hypothesis stated in terms of the former is sensitive to experimental evidence of sorts that have no bearing on hypotheses stated in terms of the latter.

The point just made is really a special instance of the general principle that covert variables, mentalistic and otherwise, are entangled in the "nomological network" in ways quite different from overt behavioral variables, verbal or otherwise. The special application of this principle reflected in the preceding paragraph concerns the way in which a mentalistic variable will typically have more or less reliable connections with a variety of observable manifestations, of which explicit verbal report constitutes one sub-type.[19] Another application of the principle concerns the fact that a given covert variable will be connected with other non-observables in ways in which no observables (including its indicators) are connected with them, and these connections will affect in a variety of ways the import of hypotheses involving that variable. For example we have a great deal of knowledge (or well-founded belief) about the conditions under which color perceptions normally occur. At a very basic level this includes all our fundamental knowledge of the psychophysics of perception; the physics of light transmission, the physiology of receptor stimulation and neural transmission, etc. Again we have a great deal of (largely unsystematized and even inexplicit) knowledge about the use to which color perception is put in cognition and action. We discriminate objects in terms of color; we assign object boundaries in terms of color gradients, etc. The same point can be made about the notion of attractiveness, though there the nomological network is undoubtedly less extensive and still less explicit. But at least we know that, ceteris paribus, a person will choose a more over a less attractive object, that attractiveness is influenced by various factors, including beliefs about the instrumental value of the object and the hedonic tone of past experiences with the object.

Our use of mentalistic variables is strongly dependent on these background entanglements. For one thing, our choice of hypotheses to take seriously is influenced by them. We would not waste time with hypotheses concerning color perception that completely ignored the influence of sensory stimulation by light waves; and the suggestion that choice varies inversely with degree of attractiveness would be ruled out of court without a hearing. Second, we are guided by this background knowledge in our choice of detection and measurement devices; we utilize verbal reports for color perception but not for degree of intelligence, because we have reason to think that one's visual perceptions are immediately available to one in a way that one's intellectual abilities are not; and we use pupillary expansion as a measure of degree of attractiveness but not of strength of achievement motivation, just to the extent that we believe that the first, but not the second, is lawfully connected with pupillary expansion. Third, the wider significance of a positive or negative outcome of a given piece of hypothesis testing will depend on the place of the hypothesis in the nomological network. Confirmation of the "disappearance of color in the Ganzfeld" hypothesis will lead us to modify simplistic ideas (if they had not already succumbed to other considerations) concerning the dependence of color perception on the physical character of the light waves stimulating the retina, and it will lead to, or reinforce, the quest for the character of the mechanisms that intervene between the transmission of the visual stimulation and the conscious color perception. Again, a given hypothesis may have been developed as a derivation from some higher level theory, or, regardless of how it was originally conceived, may have come to be taken as such a derivation. In that case, its confirmation or disconfirmation will have repercussions on the the theory and, through that, on other derivations from that theory. Thus Brehm thought of his hypothesis as a particular application of the theory of cognitive dissonance. Its confirmation increases the credibility of that theory and, through that, the credibility of a wide variety of other applications of the theory to such fields as the effects of forced compliance, voluntary exposure to information, and the effects of the disconfirmation of strongly held beliefs.

None of this is preserved under the substitution of overt verbal report dependent variables. The determinants of verbal reports are, as we have seen, quite different from the determinants of color perception; and so the constraints on the plausibility of hypotheses involving the two sorts of variables are quite different. Confirmation of hypotheses concerning the determinants of verbal reports of a given kind will stimulate the search for quite different sorts of mechanisms from the confirmation of hypotheses concerning the determinants of color perception. Confirmation of a rating behavior transformation of the Brehm hypothesis will not ramify to other derivations from the theory of cognitive dissonance, for overt ratings are not the sorts of things to which the theory applies. Thus in all these respects the verbal replacements for mentalistic hypotheses occupy a very different position in the fabric of psychological theory and research. To simply replace the latter by the former would be to rip that fabric, perhaps to shreds.

(c) I conclude that Spence's first alternative — preserving the substance of mentalistic results while reconstruing the dependent variables thereof — is doomed to failure. However interesting the revised hypotheses may be in their own right they cannot lay claim to any strong equivalence to the originals in scientific import. The substance has disappeared along with the private data.

(iv) Spence formulates his second alternative as follows:

> He also makes use of them as a basis for making inferences as to certain hypothetical or theoretical constructs which he employs.

The possibility Spence seems to have in mind here is the following. We will leave our mentalistic hypotheses as they are; "inner" variables like the character of one's visual impressions, the attitude toward an object, and so on, will continue to play the same role in our hypotheses. But instead of being regarded as observables, accessible to a special private kind of observation, they will be regarded as postulated, non-observable states or processes. We can then treat the testimony of our subjects as more or less reliable indicators or measures of various internal states, as we use a rat's crossing an obstruction as a measure of the strength of the rat's motive to get the sort of thing that is on the other side, and as we take a subject's story-telling behavior (in the TAT) as a measure of various needs.

This would seem to be a more promising way of making the best of both worlds. We countenance only publicly accessible items among our *data*; but in various ways public data may be taken to be reliable indices of internal mental states of the sorts introspectionists were accustomed to impute on the basis of supposedly direct internal observation. However on careful scrutiny we see that this alternative too is infected with fatal confusions.

Just as in the discussion of the first alternative, it will be important to get straight about the level at which the verbal utterances are conceptualized. Again the critical distinction is between reporting them in "illocutionary-act" terms ("S_1 said that the object no longer looked red"), in "locutionary-act" terms ("S_1 uttered the sentence, 'It doesn't look red now'") or in "phonetic-act" terms ("S_1 made a sequence of vocal noises of the following physical description: – "). Now if our verbal data are conceptualized at the illocutionary-act level, the assumption that they are reliable indicators of the states reported is indistinguishable from the traditional treatment of them as protocol sentences. To say that FPIPSR's (or perhaps some sub-class thereof) are reliable indicators of the inner states reported in them is precisely the assumption that is made by the "introspectionist" in taking them as protocol statements. More generally, for any statements made immediately on the basis of experience, the fact that such statements are, in general, reliable indicators for what is reported in them is a necessary and sufficient condition of justifiably using such statements as part of the empirical basis of a science. In supposing that perceptual judgments concerning rat movements made by properly qualified observers are reliable indications of such movements, we are *thereby* consenting to the validity of taking such judgments as protocol statements; *pari passu*, in supposing that FPIPSR's are reliable indicators of psychological states is *thereby* to take them as protocol statements. Thus the present "alternative" is no alternative at all. And this is the case, no matter how the assumption of reliability is justified.[20,21]

However if the behaviorist reports FPIPSR's in locutionary-act or phonetic-act terms, we can no longer claim that his position is indistinguishable from the traditional one. Let us suppose that when one of our subjects reports that the sphere now looks light gray to him, we record his response in acoustical terms, perhaps in the form of an acoustical graph. What we are now taking as a datum is that S_1 vocally produced sounds of a physical character indicated by the following graph – . We are, of course, not denying that he reported that the sphere looked light gray to him; we are simply not including any such description of his performance in the empirical data we are gathering from the experiment. As so construed, the assumption that FPIPSR's are reliable indices does not amount to an acceptance of them as protocol statements. This is because we are not now treating them as (true or false) statements at all. Hence our supposition that they are reliable indicators cannot be construed as an acceptance of them as protocol *statements*, any more than our acceptance of an increase in GSR as a reliable indicator of increased emotional arousal can be construed as an acceptance of the operation of the galvanomemter as the making of a protocol statement.

Thus in this version, Spence's second suggestion does provide a genuine alternative to introspectionism, but it does not represent a live alternative, as things now stand. In order to reconstrue mentalistic research in this fashion, we would have to have sufficient grounds for supposing, with respect to each a number of types of vocal sound production, that its occurrence (under certain physically describable circumstances) is a reliable indication of a certain inner mental state, e.g., some object looking gray to the subject. But what reasons do we, or might we, have for such a claim?

Conceptualizing FPIPSR's on the illocutionary-act level, we have reasons of a broad, rather vague sort, for supposing them to be reliable indicators of the states reported. We have learned through the ages, in ordinary daily intercourse, that if we take people's word for what they are sensing, feeling, thinking, and so on, this turns out to be a very reliable guide to their future behavior,[22] especially if we take precautions against lying and verbal misunderstanding. And by using FPIPSR's as protocol statements in research on psychophysics, perception, cognition, etc., some results have been obtained which seem to stand up fairly well in a variety of contexts other than those in which they were obtained. For example, with respect to the cues used for size constancy, results obtained in experiments relying on

FPIPSR's have been generalized with success to the perception of lower animals, where verbal reports are not available.

Now can we give an analogous argument for the reliability of certain vocally produced sound-sequences as indicators of certain inner mental states? It is conceivable that such an argument could be carried through. But we are not now in possession of the evidence on which to base such an argument. To get that evidence we would have to systematically record the acoustical qualities of verbal behavior, and then try out various hypotheses as to what kind of inner mental state a given acoustical design indicates, so as to see which of these hypotheses pan out. Since this has not been done to date we presently have no basis for taking any particular kind of acoustical design as a reliable indicator of any particular kind of mental state. Nor can we appeal to the common experience of mankind, as we did in order to support the analogous claim for FPIPSR's conceived *as* reports. For our common experience of verbal behavior is recorded at the illocutionary-act level. In our ordinary perception of speech (including psychologists' perception of their subjects' verbal reports, except where speech production is the explicit object of study) we attend to *what* the speaker is saying, the *content* of his utterance, rather than the physical characteristics of his utterance. And so what we learn from common experience concerns nomological connections of illocutionary-act types, rather than of acoustical types. Nor can we suppose that our common-sense evidence for the reliability of FPIPSR's is also evidence for the reliability of vocal sound-productions. It is true, of course, that whenever anyone (orally) issues a FPIPSR he produces some sound pattern or other. But, as we saw earlier, a given report can be issued by producing any of an indefinite number of different sound-sequences.[2,3] Moreover any one of those sound sequences can be produced to do something other than report a current mental state. For example, I can produce the sound-sequence ordinarily involved in uttering the sentence "It looks gray" in order to give an example, test a microphone, or recite a line in a poem. In none of those cases would I be reporting my current visual impressions. Thus even in the most favorable cases there is a poor fit between illocutionary acts and acoustical productions, a poor fit in both directions. This means that having established the reliability of a given type of illocutionary act, there is no automatic way of transferring that reliability to a given type of vocal sound production.

Thus we are not now in possession of any grounds for supposing that any specifiable type of sound production is a reliable indication of any specifiable inner mental state. Such grounds could conceivably be built up, but to do so would require investigations that have still to be carried out.[24] At present the behaviorist is in no position to make Spence's second move, in any form in which it is distinguishable from old-fashioned introspectionism.

(v) The upshot of the discussion in Part I is that there is no way of combining an acceptance of the behaviorist proscription of private data with the acceptance of the substance of mentalistic research. First we saw that as that research is ordinarily conceived it is committed to accepting FPIPSR's as protocol statements. Second, the sorts of reconstrual proposed by Spence will not do the job. For the only version of Spence's moves on which the character and import of the research was not radically changed turned out to be committed after all to accepting FPIPSR's as protocol statements along with reports of public observations. Therefore we are compelled to regard the popular "we can have it both ways" position, set forth by Spence, as sheer bluff. We cannot enjoy the fruits of theorizing and experimentation carried on within a mentalistic framework and at the same time observe a ban on private data. The only coherent program for avoiding private data would be one which resolutely turned its back on all the mentalistic research of the past and started with a clean slate to explore (in physicalistic terms) the determinants of overt behavior. This heroic course abandons any hope of preserving the results of introspectionist psychology, and so it has not found as much favor as the "in principle it could all be reinterpreted" move.

PART II

If the conclusions of Part I are justified, something will have to go. Either a large part of the results of experimental psychology will have to be thrown out and research programs drastically revamped, or else the ban on "private" data will have to be lifted. It may seem to many a desperate choice. Either admit that most of psychology has been built on sand or abandon the scientific pretensions of the discipline. However on closer scrutiny the second horn of the dilemma turns out to be a paper horn. If we carefully scrutinize the arguments that led to the rejection of introspective data, and still give it such support as it enjoys, we will see that this rejection is not really supported by any fundamental requirements of scientific method. We can use introspective data, subject to basically the same safeguards as are needed for any other kind of data, and enjoy scientific respectability at the same time.

(i) The first step in restoring the legitimate rights of FPIPSR's is to separate out the arguments against the precise claim we are defending — viz., that FPIPSR's can, under suitable conditions, be accepted as protocol statements — from arguments against other features of structuralist psychology that are not relevant to the present issue. The structuralist psychology of Wundt and Titchener was a complex body of thought, rich in philosophical and methodological presuppositions as well as in detailed empirical investigations. The commitment to introspection as the specifically psychological way of gathering data was only one of its distinctive features. Criticism of such of its principles as are not logically tied to the reliance on introspection are not relevant to the present discussion. These principles include the "atomistic" view of consciousness as compunded out of simple, unanalysable conscious elements, and the conception of psychology as essentially the analysis of consciousness. Once could recognize the protocol status of (some) FPIPSR's while rejecting both of these principles. The same must be said about the entanglement of our issue with metaphysical disputes between dualists and materialists. It is easy to suppose that a reliance on introspection carries with it a commitment to an anti-materialist conception of the mind. It seems plausible to suppose that if and only if man is a purely physical system, one should be able to investigate him with the same observational tools that are used for any other physical system, whereas if and only if he is of a radically different nature some radically different kind of observation is needed. And in fact introspection has often been attacked in the name of a materialist conception of man. However there is really no logical connection between the two issues. It is not at all inconsistent with materialism to hold that a human being has some special observational access to certain of his states and processes, which other people, if left to their own devices, could learn about only through complicated inferences. A materialist will, of course, hold that this self-observation is carried out by the use of some kind of physical mechanism, but there is nothing in that supposition to inhibit us from using FPIPSR's as protocol statements.[25]

But even within that part of the behaviorist-structuralist debate that had to do specifically with introspection there are issues that must be distinguished from the one with which we are presently concerned. These are issues that stem from features of structuralist conceptions of and claims for introspection that are not duplicated in our conception of FPIPSR's. Let us briefly survey these.

First introspection was traditionally conceived as a kind of observation, differing from sense-observation in the nature of its objects and in the fact that it did not employ the external senses, but otherwise sharing in the generic characteristics of observation. It has distinguishable objects, which one can single out and to which one can pay more or less attention. It is under one's control; one can initiate it and shut it off; one can devote more or less of one's attention and energy to it; one can perform it more or less carefully, systematically, and thoroughly. It is no part of my concept of a FPIPSR that it stem from introspection so conceived. Therefore one can allow FPIPSR's to count as protocol statements, without rendering himself vulnerable to the following sorts of criticisms, all of which are specifically directed against the conception of introspection as a kind of inner observation.

(1) Directing one's attention on a conscious state inevitably changes its character. Hence by introspection we cannot find out what conscious states are like when they are not being introspected.

(2) One cannot simultaneously be experiencing a conscious state *and* introspecting it. Hence the most we can do is *remember* conscious states just after they happen.

(3) Philosophical arguments to the effect that one's awareness of his sensations and feelings does not exhibit the essential features of observation, e.g., Ryle's argument to the effect that it does not make sense to speak of having a sensation more or less carefully or methodically.

Second, introspection has been thought of as both infallible and omniscient. It has been supposed that one could not possibly misperceive his own conscious states, and that it is equally impossible for one to fail to notice one of his conscious states, provided he puts his mind to it. These claims are alluded to by Spence in the above passage, when he says, "The introspectionist, it should be recalled, assumed a strict one-to-one relationship between the verbal responses of his subjects and the inner mental processes." Such claims have been made by philosophers rather than by experimental psychologists. It is because the structuralist psychologists recognized the possibility of mistakes in introspection that they insisted on careful training for experimental subjects before their introspective reports could be used as evidence. But whatever the history of the subject, it is clear that one does not have to claim infallibility for FPIPSR's in order to count them as protocol statements; and we are making no such claim for them.

Third, structuralists typically regarded introspection as capable of carrying the entire observational load for psychology (apart from data about the external environment). If the mind is the field of consciousness, and if psychology is the science of the mind, then introspection should provide us with all the specifically psychological data we need. The defender of FPIPSR's need not make any such claim, and I am not at all disposed to. Hence the following criticisms are not relevant to my thesis.

(1) Since the mind is more than consciousness, introspection alone is not sufficient.

(2) Psychology is the science of behavior as well as (or instead of) the science of the mind.

(3) If every act of introspection is itself introspectable we are faced with the spectre of an infinite regress. If not, there are some mental acts that cannot be observed by introspection.

Lastly, the fall of structuralism was largely brought about by its failure to resolve certain specific problems in which a lot of research had been invested, particularly the question of whether there is imageless thought, and the analysis of feelings into sensations. Critics pointed out, with justice, that consistent results were not forthcoming on these questions, and that introspective reports bearing on these issues tended to vary from laboratory to laboratory in a way that closely reflected the theoretical prepossessions of the director of the laboratory in question. Now my thesis is that there is no general methodological bar to the utilization of FPIPSR's as protocol statements. I am not thereby committed to their usability in any particular area, and hence not in those areas in which they proved themselves sadly lacking. One can recognize that human beings have the capacity to make reliable non-inferential reports concerning some of their states while at the same time recognizing that there are other states with respect to which they are not so endowed. In the investigations just alluded to the structuralists undoubtedly made unreasonable demands on introspection. But the fact that there are some tasks it cannot perform does not imply that it is completely inept. More recently the unreliability of self-ascriptions of personality traits and the modes of self-deception uncovered by psychoanalysis have been invoked as reasons for a general rejection of self-attributions in psychology. But again one should be alive to the possibility that there are other areas, e.g., visual appearances, in which immediate self-attribution is much more reliable, perhaps at least as reliable as careful observation of the physical environment.

(ii) To return to the issue with which we are specifically concerned, what arguments are there that are both worthy of serious consideration *and* specifically opposed to the thesis that FPIPSR's, in our sense of that term, can function as protocol statements? It seems to me that there is one and only one such argument, viz., that there is no public check on the accuracy of FPIPSR's. Where a report of observation concerns an external, physically described state of affairs, there is always the possibility of one or more other investigators observing the same situation and providing independent testimony. For example, in investigating the effects of periodicity of reinforcement, the experimenter notes on which trials his subjects enter the left arm of a T-maze. Perhaps he is recording the trials in solitude, and no other person is actually in a position to check up on his reports. Even so *if* we had any reason to question his accuracy we could obtain a number of independent reports on the transactions. The credentials of any particular report could be assessed in terms of its agreement or lack of agreement with others. With FPIPSR's, on the other hand, these moves are ruled out. By definition a FPIPSR has to do with a state of affairs concerning which at most one person can give an immediate, non-inferential report. Hence where it is a matter, e.g., of how the sphere looks *to me* at the moment, there can be no question, as there was with the movements of the rat in the T-maze, of bringing in other observers to serve as a check on the accuracy of my reports. In the nature of the case it is impossible for anyone else to "take a look" at numerically the same state of affairs. Since FPIPSR's remain an "unknown quantity" we are not justified in accepting any of them as providing empirical data. Let us call this the "consensual corroboration" argument.

Before coming to terms with the argument I want to distinguish it from related points which are often made on this score, but which can be quickly dismissed as based on confusions. For example, we hear a lot from methodologically minded psychologists about "inter-subjective consistency" and "inter-observer reliability," and the credentials of FPIPSR's are often impugned on the grounds that they do not satisfy these requirements. But inter-observer consistency is a valid requirement only if the observers in question are observing the same thing. If A is observing rat R_1 in maze M_1 at time t_1, and B is observing R_2 in M_2 at t_2, there is no special reason to expect their reports to agree; we are obviously unjustified in laying it down as a conditions of acceptability of rat-behavior observation reports that the report of one rat's behavior agree with the report of another rat's behavior. But this is just the situation with respect to FPIPSR's. By the nature of the case two different subjects issuing FPIPSR's are reporting the states or reactions of two different persons. So that even if the physical environment of the two subjects is, so far as we can tell identical, there is no general reason to suppose that the facts reported by the subjects are the same. There is certainly no general methodological reason for thinking that all persons will perceive a certain physical state of affairs in the same way, or that they will all find a given object equally desirable (even given identical experimental "treatments"). Indeed, there is every reason to expect a significant degree of individual variation in these matters. Thus the mere fact that introspective reports vary under identical physical conditions[26] gives us absolutely no reason to question such reports. For all methodology can tell us, such variations are to be explained in terms of actual differences in the facts reported (due to the differences in the subjects), rather than in terms of the unreliability of the observation process. Thus if any criticisms along these general lines is to be constructed, it will have to be that there is no way of choosing between the above hypotheses. That is, if two persons, under identical experimental conditions, report different perceptions, there is no way of deciding whether they are in fact seeing the situation differently (for reasons to be further explored), or whether one or both is incorrectly reporting his perceptions. But this charge can only be supported by pointing out that it is in principle impossible for another observer to report the *same* perception, and hence we are debarred from determining whether it is inaccuracy of observation that is responsible for the divergence. Thus the argument from lack of inter-observer agreement, once purged of confusion, readily reduces to the argument already presented.

The consensual corroboration argument certainly appears to have force, and it has convinced a number of thinkers. Nevertheless, I believe it to be invalidated by a very simple mistake, viz., supposing that the accuracy of a putative protocol statement can be assessed *only* by obtaining other experiental reports of the same state of affairs. Let me elaborate.

It *is* a basic methodological requirement on putative protocol statements that there be some way of determining their accuracy. This requirement stems from the fact that experiential reports can be mistaken. If every report of experience carried an infallible warrant of accuracy on its face, we could, without more ado and with complete assurance, welcome any such report into the empirical basis of science. But such is not the case. It is not at all uncommon for persons to confidently assert P as a report of their experience, where P is not the case. This could be because the person is in an unfavorable position to determine whether P is the case, because of a malfunctioning of his perceptual apparatus, because of motivational distortions, and so on. Now we certainly want science to rest on fact and not fancy. This means that we must have some basis for choosing, from among putative protocol statements, those that are most likely to be accurate. Hence it is crucially important to have some way of determining the accuracy of experiential reports.[27]

The possibility of gathering testimony from other observers of the same state of affairs is a methodological desideratum only because it provides one way of satisfying the above requirement. The fact that this is, the *only* source of its methodological desirability is shown by the fact that consensual corroboration is thought to be methodologically desirable only when we are justified in supposing that where a number of different observers all agree in their report, we can have more assurance that the content of the report is wholly determined by the objective facts of the matter than we can where a single report is all we have to go on. In cases where we are not justified in making this assumption, e.g., cases in which the most reasonable assumption is that the consensus is due to mass hysteria, we do not feel that the possibility of consulting other observers confers any methodological advantage.

But if the possibility of consensual corroboration is desirable only because it enables us to check the accuracy of an experiential report, then we are justified in making it a *necessary* condition of methodological respectability only if consulting other obseervers of the *same fact* is the only way of carrying out such a check. But in general this is not the case. This point can be illustrated with uncontroversial examples drawn from statements for which the testimony of other observers is *also* in principle available. Let us suppose that I am the only person on a lonely Pacific atoll, which happens to be in a rocket testing area. One day I see a rocket whizzing directly overhead, moving in a northeasterly direction. I glance at my watch and note that is is 11:23 a.m. Of course rocket movements are among the sorts of things that are open to public observation, and so it is clear that there is no impossibility in there having been other observers on the spot to serve as a check on my perceptual judgment that a rocket passed over Iwani at 11:23 a.m. of that day, moving in a northeasterly direction. (Call this judgment 'R'.) Nevertheless in fact there were no other observers on the spot. Does this mean that no one can investigate the accuracy of R? By no means. Even though no one else directly observed the rocket passing over Iwani, there is a wealth of other information that bears on the question. The rocket was fired in such a way as to take a certain course (including a pass over Iwani at 11:23 a.m.), and in the absence of evidence to the contrary it can be assumed that it proceeded according to plan.

Furthermore it was sighted by numerous people at various other places and times in such a way that by filling in the gaps, in the light of our general knowledge of rockets plus particular facts about this firing, we plot a trajectory that includes a passage over Iwani at 11:23 a.m. Finally, tracking stations kept tab on the course of the rocket by means of radar and other such devices, so that by an elaborate process of inference from instrumental readings, it could be concluded that R was correct. And all this in the (*de facto*) absence of other persons who observed just the same situation as I.

What this case reveals is that there are many ways of checking the accuracy of a protocol statement, other than by additional observers "taking a look" at just the situation reported in that statement. This being the case, we cannot assume, without more ado, that none of these other ways are available in the cases of statements, like FPIPSR's, where the possibility of other observers of the same fact is ruled out. Hence the consensual verification argument fails. Corroboration from other observers of the same state of affairs is not the only way of checking up on the accuracy of a report. Hence we cannot argue from the impossibility of there being other obsevers of a state of affairs, S, to the impossibility of determining the accuracy of an experiential report concerning S. And we have seen that such an implication constitutes the only ground for making the possibility of consensual corroboration a methodological requirement.

(iii) The above considerations, I take it, are sufficient to dispose of any general methodological arguments against the claim that FPIPSR's are respectable candidates for protocol statements, but of course they do not constitute any positive support for that claim. They merely open the way for such support. I have already argued that FPIPSR's are in fact treated as protocol statements in many areas of psychological research that are very much going concerns. This creates a prior presumption in favor of the principle; the burden of proof, one might say, falls on the opposition, and if the opposing arguments fail, that settles the matter. However I do not want to confine myself to scoring debaters' points. I have agreed that the possibility of an independent check on accuracy is a reasonable demand to put on putative protocol statements, and even though we have disposed of general methodological reasons for supposing the FPIPSR's are insusceptible to such a check, one may still wonder whether they are in fact so susceptible and if so how. I cannot embark on a thorough discussion at the tag-end of this paper, but I shall just indicate how, as I see it, one can investigate the accuracy of FPIPSR's. The matter has already been touched on above (p. 268f), but it would be well to set the topic in a wider perspective.

From the rocket case we can derive the following highly general statement of what is involved in checking out a putative protocol statement *without* gathering testimony from other observers of just the same situation. Where we have knowledge, or well-founded belief, concerning lawlike connections of facts of the sort reported with facts of other sorts, we can use this knowledge (together with knowledge of other particular facts) to determine what other observable facts would be expected if the putative protocol statements were accurate; we can then use observation to determine whether these other facts are as predicted. In the rocket case the general knowledge included that involved in our ability to trace the path of a rocket, given facts about the initial propulsion, direction, atmospheric conditions, etc., as well as the general knowledge involved in our ability to interpret and integrate radar signals. Knowledge of other particular facts included knowledge of the details of the firing, intermediate sightings, and the termination of the flight. All of this gave us a substantial basis for arriving at a judgment as to whether the rocket was in fact passing over Iwani at 11:23 a.m.[28]

So the question of whether it is possible to check the accuracy of a particular FPIPSR boils down to the closely interrelated questions: (1) Do we have knowledge of general connections between mental facts of the sorts reported in FPIPSR's and other sorts of facts, and (2) can we get at these other sorts of facts, directly or otherwise?

Of course, these questions, resolutely pursued, would take us into every corner of psychology and the philosophy of mind. Here I can do no more than to remind the reader that our grasp of such matters, while grievously lacking in completeness, precision, explicitness, and certainty, is by no means negligible, and hence that we are by no means helpless when it comes to investigating the accuracy of a given FPIPSR. I shall direct my remarks to the kinds of mental states involved in our two research examples.

With respect to "conative dispositions" like desires, aversions, likes, dislikes, attitudes, senses of desirability, etc., there are obvious connections with overt behavior. A person can be expected to choose things he likes, reject things he dislikes, support causes he favors but not those he disfavors, and so on. To be more specific, take an FPIPSR from the Brehm experiment. Let's say that one of the subjects indicates that she finds the toaster "very desirable" but the stop watch "not very desirable." If this report accurately represents here disposition vis-a-vis these objects, then if she is placed in a situation where she has to choose between them, we would expect her to choose the former.[29] Such actual choice behavior could serve as a check on the accuracy of the FPIPSR. To be sure, these connections with overt behavior are invariably mediated by a number of other factors. If one really does favor withdrawal from Vietnam he can be expected to support petition drives aimed at that goal, *provided he does not have stronger reasons for refraining* (e.g., fears that he will be considered "unpatriotic"), and provided there are not more pressing concerns that leave him no time for this, and. . . . One who has a strong desire to become a dean will do things he believes will increase his chances for such a position, unless he is inhibited by scruples or fears of the consequences, etc.; and just what he will do depends on what he believes to be most likely to be effective. The fact that a given conative disposition is related to overt behavior not in any direct or simple manner, but only in the context of various other factors, psychological and situational, has a number of important consequences. The most significant one for our present discussion is that no simple demonstration of the inaccuracy of a first-person report is possible. Since the attribution of a conative disposition yields predictions about overt behavior only together with assumptions about other psychological factors, a failure of the prediction might conceivably be explained in a number of different ways (any one of the factors involved in the derivation might be responsible). Let's say that Jones reports being strongly opposed to the Vietnam war, but then fails to support any anti-war drives. Does that show that his report is inaccurate? Not necessarily. The failure to act might be due to one of the other factors involved, e.g., a fear that so acting would jeopardize his relations with influential people in the community. Again, if we consider Brehm's subjects who made choices other than what their ratings would have indicated, their behaviour can be viewed as a disconfirmation of their FPIPSR, but it *might* also be explainable in terms of a shift in attitude between rating and choice, or, conceivably, by a belief that they would "look better" if they chose the item they did in fact choose. Thus we can be justified in concluding that the FPIPSR is mistaken, only to the extent that we are justified in supposing that the inaccuracy of the FPIPSR gives a better explanation of the discrepant behavior than any other alternative. This complicates the matter, but it does not leave us helpless; for we can often be justified (using reasonable standards) in ruling out explanations other than the falsity of the FPIPSR's. In the Brehm case, e.g., it is implausible (in the light of what we know about such matters) to suppose that preferences would have shifted that quickly, and there was no reason to suppose that anything extraneous was motivating the choice. Therefore the most plausible explanation is that the FPIPSR's incorrectly represented the preference hierarchy.

A crucial feature of the above procedure is that in investigating the accuracy of any given FPIPSR we are making use of a variety of assumptions about other psychological states of the individual, as well as assuming general psychological principles; and it seems clear that we have to rely on other FPIPSR's if we are to justify such assumptions. Does this make the procedure viciously circular? It will undoubtedly seem vicious to one with a faith in the possibility of a strictly "linear" procedure in science, in which we start with unquestionable data, moving by logical steps from stage to stage, never utilizing any premises except those that are either unquestionable in themselves or have already been established by this procedure. Such a person might demand, as a condition of accepting any FPIPSR's, that any given statement of this sort be established without using any devices except public data and other facts and principles that had already been inductively established on the basis of public data. Recent work in the philosophy of science has shown, conclusively in my opinion, that the logic of science cannot be represented in so simple a fashion. Things are much messier than that. At any given point we have to make use of assumptions that have not been derived in this fashion from any absolute rock-bottom. Any particular assumption can subsequently be tested, but only against the background of other more or less dubitable assumptions. Applied to our present concerns, this means that any given FPIPSR can be criticized in the light of a variety of principles and facts we are currently disposed to accept (including beliefs about other inner mental states we have gotten at through FPIPSR's), where any of these latter may be similarly questioned in turn.

Thus, given reasonable standards for scientific evidence, it is possible to gather evidence for or against the accuracy of a given first-person report of a conative disposition, although we should not expect to decisively settle the question in every instance. This is sufficient for any reasonable methodological requirement of testability.[30]

Basically the same points can be made concerning such things as visual impressions, though the connections with overt behavior are still more tenuous. Even in the sphere of conative dispositions it is possible for a person to have a given desire, liking, or attitude and not do anything about it; but this is much more prevalent when we come to consider such items as sensations, perceptions, and mental images. Many of these "conscious contents" are not reflected in behavior in any obvious way. However the crucial point about perception, at any rate, is that it always has the potential function of guiding our behavior. If we are to carry out our aims we need a constant stream of perceptual information concerning the physical environment within which our endeavors take place. Although many of our perceptions are not in fact made use of in this way, they all are inherently fitted to play this role. It is undoubtedly to our advantage that our perceptual apparatus provides us with a *superabundance* of information, leaving us to pick from among this what has a bearing on the activities of the moment. Thus we can state, as a general law-like truth about perception, that when a certain feature of the environment is relevant to a person's aims at the moment, he will act in one way rather than another, depending on how he perceives that feature. If my dominant aim at the moment is to walk safely across the street, then my precise pattern of movements will be influenced by the way in which I perceive the traffic in the street. Thus if we want to institute a check on the accuracy of a certain person's report of his color perceptions, we simply find, or institute, a situation in which the color of some object is relevant to his current aims; driving where there are traffic lights would be a simple example. We get him to say what color a given object appears to him to have and we then determine whether he acts as we would expect him to if he does perceive the object in that way. For example, we get him, while driving, to tell us what color a traffic light in front of us appears to him to have, and then (assuming that part of his current aim is to obey the law) we note whether he stops or continues. This technique will not enable us to institute a direct check on any sensory impression report taken at random, for in many particular cases the character of one's visual impressions will not have any foreseeable influence on behavior. In the context of the Ganzfeld experiment, e.g., whether one sees the sphere as colored or not after three minutes could not be expected to influence any publicly observable behavior, except the report whose accuracy is being assessed. However the point is that for any *sort* of sensory-impression report, e.g., reports on color impressions, one can find or institute situations in which the character of one's sensations can be expected to influence behavior in predictable ways. One is thereby in a position to investigate the credentials of each type of putative protocol statement, and, furthermore, to investigate the accuracy of a given subject with respect to statements of that type. As we pointed out in footnote 27 this is quite sufficient for satisfying the basic methodological requirement of testability.

(iv) The upshot of the discussion in Part II is that, first appearances to the contrary, the insusceptibility of FPIPSR's to corroboration by other observers of the same state of affairs constitutes no bar to accepting them, or some of them, as protocol statements. Despite the impossibility of anyone else's experiencing the same thing, the accuracy of such statements can be investigated in a variety of ways, and that is all that one can reasonably require on methodological grounds. Hence the collapse (documented in Part I) of behaviorist attempts to preserve mentalistic research while rejecting private data, presents no cause for despair. Since one can accept private data, while being as scientific as one's behaviorist counterparts, the attempt need never have been made. Its failure is the dénouement of a farce, not a tragedy.

Douglass College, Rutgers University

NOTES

[1] From 'The Methods and Postulates of Behaviorism', *Psychol. Rev.*, 1948, pp. 67–78.
[2] If anyone is worried about the boundaries of this term, it can be dropped without affecting the discriminatory power of the definition.
[3] I am not restricting "inference" to "conscious inference," in which the person consciously formulates premises and conclusions.
[4] For a discussion of epistemological problems concerning one's reports of one's contemporaneous psychological states see my article, 'Varieties of Privileged Access', *American Philosophical Quarterly*, 8, No. 3, July, 1971.
[5] Strictly speaking, the considerations we have presented do not require that a *line* be drawn. Our claims as to what is required to make science "empirical" are quite compatible with recognizing all statements to fall within the class of protocol statements. However a proposal to do so would be wildly implausible.
[6] More precisely, "not to be treated as *yielding* protocol statements." Precision requires this refinement because we have introduced the notion of FPIPSR as a category of datable events, each of which is a particular verbal utterance, whereas we are thinking of a protocol statement as a sort of common possession of the scientific community, which, once acquired, is there to be built on and to be used in the testing of various hypotheses. A permanent possession that can be used repeatedly cannot be identified with a momentary event. What we are asking is whether the *statement that is made* when one issues FPIPSR can thereby acquire the status of a protocol statement in the scientific community. For simplicity of exposition, we shall ignore this refinement and speak as if FPIPSR's and protocol statements were nearly enough of the same category to be identifiable with each other.
[7] Thus our issue could be formulated simply as "Are FPIPSR's to be treated as

statements?" IF they are treated as statements at all, it will presumably be as protocol statements. For further discussion of this point, see pp. 85–89. (This volume, pp. 267–272).

[8] *Exper. Psychol.* 41 (1951), 153–159.

[9] *Abnorm. Soc. Psychol.* 52 (1956), 384–389.

[10] A positive number indicates an average percentage increase in desirability rating; a negative number an average percentage decrease.

[11] *Loc. cit.*

[12] It may be that Spence did not have in mind this way of reconstruing mentalistic research, but was only pointing out that one can *also* try to discover laws governing verbal behavior. There is no objection to such an enterprise, of course; but on this interpretation Spence's remarks are not germane ot the question of how the behaviorist might retain research that is (originally) conceived as dealing with inner mental states. In any event, however Spence is to be interpreted, this "first possibility" is often taken seriously by behaviorists, and it is what we wish to consider here.

[13] This association is made explicit by Spence in the article from which we have been quoting.

[14] By a curious irony the argument has now taken such a turn that the behaviorist is defending the position attributed by Spence to the introspectionist, "a strict one-to-one relationship between the verbal responses of his subjects and the inner mental processes." For in saying that the verbal reports are under functional control of the same variables as the inner processes reported, the behaviorist would appear to be assuming such a correspondence. And his critic, in emphasizing functional divergencies between the two, would, correlatively, appear to be denying it. Actually this issue is a red herring. In practice, both introspectionist and behaviorist will assume that verbal reports can be depended on to reflect "inner mental processes" when and only when certain further conditions (of the sort mentioned below) are satisfied. Neither party has to assume that such reports are infallible, or even that they are generally reliable when such conditions have not been satisfied.

[15] For an explanation of the relevant sense of this term see my *Philosophy of Language*, Englewood Cliffs, N.J.: Prentice-Hall, 1964, pp. 34–36.

[16] This lower level of conceptualization is, of course, the one suggested by Spence's statement: "He accepts verbal response as just one more form of behavior and he proposes to use this type of data in exactly the same manner as he does other types of behavior variables."

[17] Moreover it seems clear that such a formulation would dissipate the original intent of the research. The idea was that the antecendent conditions specified would give rise to a distinctive sort of result, not just to some result or other out of a large set of alternatives.

[18] The relation between a formulation in terms of a statement with a certain content and one in terms of utterances described phonemically or phonetically, is parallel to the distinction between the description of non-verbal behavior in terms of goals aimed at, and a description in terms of muscle movements. Here too we often have reason to accept a hypothesis as to the conditions under which an organism will do something to bring about a certain result, e.g., take a right turn in a *T*-maze; but since there are an indefinite number of muscle-movement sequences that can be used to bring that about, we cannot simply restate the same hypothesis with the dependent variable specified in

terms of specific muscle movements. For an illuminating discussion of this issue, see C. W. Taylor, *The Explanation of Behavior*, London Routledge and Kegan Paul, 1964, Ch. VII.

[19] This is one of the considerations that proved fatal to the operationalist attempt to identify a given covert variable with a particular overt indicator.

[20] For some discussion of what is involved in assessing the reliability of putative protocol statements, see Part II.

[21] There would be a genuine alternative here only if Spence were correct in foisting an infallibility principle onto the introspectionist. For the behaviorist alternative presently under consideration does not involve the supposition that FPIPSR's are infallible indicators of inner mental states. However, as pointed out earlier, it is a gross historical mistake to suppose that any such claim was embodied in the introspectionist position. Indeed, whatever the tenets of any particular school of psychology, the basic question is whether an admission of fallibility prevents one from taking a certain class of statements as protocol statements; and the behaviorist's own practice with physical object reports shows that he cannot maintain that it does.

[22] The use of this information is, of course, severely limited by our rather scanty stock of general principles relating psychological states to behavior, and when we consider the extent of our inability to predict another person's behavior even after he has told us a great deal about his perceptions, attitudes, beliefs, etc., a large share of this impotence must be attributed to the poverty of the nomological network in which we imbed the particular information. The above point could be more precisely stated as follows: Considering the resources we have for using information about particular mental states in the prediction of behavior, it seems clear that by using information contained FPIPSR's we are much better off than we otherwise would be.

[23] Note that we cannot rule out this version of Spence's second alternative just on the grounds that there are an indefinite number of physically distinct ways of making a given report. This point was fatal to the first alternative, for that involved taking overt utterances as dependent variables in laws, and so the indefinite variety just alluded to would infect the body of nomological principles. However on the second alternative the behaviorist is thinking of overt utterances as indicators of dependent variables, rather than as themselves figuring in laws. And so the open-endedness (of the list of physical devices for making a given report) can be regarded as just a special case of the general phenomenon of an indefinite variety of indicators for a given covert variable, a phenomenon exhibited throughout science.

[24] It would seem that the systematic development of principles relating vocal sound types to inner mental states would involve pushing psycholinguistics far beyond its present condition.

[25] And it is also consistent with dualism to hold that men have no such privileged access, though the belief in privileged access has been (misguidedly, I believe) one of the main supports for dualism.

[26] Identical as far as the gross external aspects are concerned. If two subjects reported different visual impressions when the physical situations were identical in every respect, including the states of the two central nervous systems, that would be another matter. Needless to say, at present we are not in a position to determine in detail sameness or difference of the states of the central nervous system. If we were, we would presumably not have to rely on FPIPSR's.

[27] Strictly speaking, it is not essential that we be able to run an independent check on each individual alleged protocol statement. It would be sufficient to determine for each class of statements (e.g., perceptual reports of the color of physical objects), the conditions under which a statement of that class is likely to be accurate, when issued as a report of experience. And for this it would be (in practice) necessary to be able to make an individual determination of accuracy for *some* putative protocol statements of that class. Hence we shall understand the principle to require that (given a suitable classification of possible protocol statements) any candidate for protocolhood must belong to a class of statements such that it is possible to determine the accuracy of members of that class.

[28] In a more profound view of the matter, corroboration by other witnesses at the same time and place would not be contrasted with the kinds of checks broadly described above, but rather would be construed as a special case of the above. In order for A's perceptual report to confirm or disconfirm B's, certain sorts of conditions have to be satisfied; they have to be in the same place at the same time, looking in the same direction; B's sensory apparatus has to be in good working order, he must be in a "normal" condition otherwise in relevant respects; he must be a qualified observer for the sort of object under consideration, and so on. (These requirements have to be filled in differently for different kinds of subject-matters. One might be qualified to rule on whether there were trees in front of one, but not on whether a rocket was passing overhead.) In picking out just these conditions we are relying on a complex network of general knowledge concerning perception and the ways in which it is or is not determined, in various respects, by the objects perceived. This means that when I take B's perceptual report of the same fact to corroborate A's original report, my reasons for doing so involve the assumption of nomological relations between the two reports, just as much as do my reasons for taking reports of the intial firing of the rocket as corroborating its sighting over Iwani Island.

[29] The Brehm experiment did place most of the subjects in such a choice situation with respect to a pair of the items. Some subjects did make choices that were not in line with the original ratings, and these subjects were excluded from the final results.

[30] A thorough treatment would also include consideration of checks that are based on connections of a given conative disposition with other kinds of (non-publicly observable) psychological states. For example, it is generally true that when x seems attractive to a person, P, P believes that x will satisfy some of his desires, P will feel disappointed if he fails to get an x he had expected to get, and P will be opposed to proposals to make x's illegal. Insofar as we are in a position to determine whether these further psychological statements are true of P, this provides us further resources for checking self-attributions.

BIBLIOGRAPHY

Allport, G. W., 'The Eidetic Image and the After-Image', *Am. J. Psychol.* **40** (1928), 418–425.
Anderson, J. R. and Bower, G. H., *Human Associative Memory*, Washington, D.C.: Winston, 1973.
Anderson, R. C. and McGaw, B., 'On the Representation of Meaning of General Terms', *J. Exp. Psychol.* **101** (1973), 301–306.
Arbib, M. A., 'Automata Theory as an Abstract Boundary Condition for the Study of Information Processing in the Nervous System,' in K. N. Leibovic (ed.), *Information Processing in the Nervous System*, New York: Springer-Verlag, 1969.
Arbib, M. A., *The Metaphorical Brain*, New York: Wiley, 1972.
Armstrong, D. M., *A Materialist Theory of Mind*, New York: Humanities Press, 1968.
Armstrong, D. M., 'Materialism, Properties and Predicates', *The Monist* **56** (1972), 163–174.
Arnheim, R., 'The Gestalt Theory of Expression', *Psychol. Rev.* **56** (1949), 156–171.
Arnheim, R., *Visual Thinking*, Berkeley: University of California Press, 1969.
Attneave, F., 'Representations of Physical Space', in A. W. Melton and E. Martin (eds.), *Coding Processes in Human Memory*, Washington, D.C.: Winston, 1972.
Atwood, G., 'An Experimental Study of Visual Imagination and Memory', *Cognitive Psychol.* **2** (1971), 290–299.
Aune, B., *Knowledge, Mind, and Nature: An Introduction to Theory of Knowledge and the Philosophy of Mind*, New York: Random House, 1967.
Ayer, A. J., *The Problem of Knowledge*, Harmondsworth: Penguin, 1956.
Bahrick, H. P. and Boucher, B., 'Retention of Visual and Verbal Codes of the Same Stimuli', *J. Exp. Psychol.* **78** (1968), 417–422.
Baylor, G. W., *A Treatise on the Mind's Eye:* An Empirical Investigation of Visual Mental Imagery, (Doctoral dissertation, Carnegie-Mellon University) Ann Arbor, Mich.: University Microfilms, 1972. No. 72–12,699.
Baylor, G. W., 'Program and Protocol Analysis on a Mental Imagery Task', in *Proceedings Second International Joint Conference on Artificial Intelligence*, London: The British Computer Society, 1971, pp. 218–237.
Baylor, G. W., 'Modelling the Mind's Eye', in A. Elithorn and D. Jones (eds.), *Artificial and Human Thinking*, Amsterdam: Elsevier, 1973, pp. 283–297.
Bekesy, G. von, 'Synchronism of Neural Discharges and Their Demultiplication in Pitch Perception on the Skin and in Hearing', *J. Acoust. Soc. Am.* **31** (1959), 338–349.
Bekesy, G. von, *Experiments in Hearing*, New York: McGraw-Hill Book Co., 1960.
Bekesy, G. von, *Sensory Inhibition*, Princeton: Princeton University Press, 1967.
Benevento, L. A., Creutzfeldt, O. D., and Kuhnt, U., 'Significance of Intracortical Inhibition in the Visual Cortex: Data and Model', *Nat. New Biol.* **238** (1972), 124–125.

Bernstein, N., *The Co-ordination and Regulation of Movements*, New York: Pergamon Press, 1967.
Bernstein, R. J., 'The Challenge of Scientific Materialism', *Internat. Philosophic. Quart.* 8 (1968), 252-275.
Bexton, W. H., Heron, W., and Scott, T. H., 'Effects of Decreased Variation in the Sensory Environment', *Canad. J. Psychol.* 8 (1954), 70-76.
Bindra, D., 'The Problem of Subjective Experience: Puzzlement on Reading R. W. Sperry's "A Modified Concept of Consciousness"', *Psychol. Rev.* 77 (1970), 581-584.
Binet, A., *L'Etude Experimentale de l'Intelligence*, Paris: Schleicher, 1903.
Blakemore, C., 'Developmental Factors in the Formation of Feature Extracting Neurons', in F. O. Schmitt and F. G. Worden (eds.), *The Neurosciences Third Study Program*, Cambridge: The MIT Press, 1974, pp. 105-113.
Blakemore, C. and Campbell, F. W., 'On the Existence of Neurones in the Human Visual System Selectively Sensitive to the Orientation and Size of Retinal Images', *J. Physiol.* 203 (1969), 237-260.
Bohm, D., *Causality and Chance in Modern Physics*, Philadelphia: University of Pennsylvania Press, 1957.
Bohm, D., *The Special Theory of Relativity*, New York: W. A. Benjamin, Inc., 1965.
Bohm, D., 'Quantum Theory as an Indication of a New Order in Physics. Part A. The Development of New Orders as Shown Through the History of Physics', *Foundat. Phys.* 1 (1971), 359-381.
Bohm, D., 'Quantum Theory as an Indication of a New Order in Physics. Part B. Implicate and Explicate Order in Physical Law', *Foundat. Phys.* 3 (1973), 139-168.
Bohr, N., *Atomic Physics and Human Knowledge*, New York: Vintage Press, 1966.
Bousfield, W. A. and Cohen, B. H., 'The Occurrence of Clustering in the Recall of Randomly Arranged Words of Different Frequencies of Usage', *J. General Psychol.* 52 (1955), 83-95.
Bower, G. H., 'Mental Imagery and Associative Learning', in L. W. Gregg (ed.), *Cognition in Learning and Memory*, New York: Wiley, 1972.
Brainerd, C. J., 'Imagery as a Dependent Variable', *Am. Psychol.* 26 (1971), 599-600.
Bransford, J. D. and Franks, J. J., 'The Abstraction of Linguistic Ideas', *Cognitive Psychol.* 2 (1971), 331-350.
Bransford, J. D., Barclay, J. R., and Franks, J. J., 'Sentence Memory: A Constructive Versus Integrative Approach', *Cognitive Psychol.* 3 (1972), 193-209.
Bransford, J. D. and Johnson, M. K., 'Consideration of Some Problems of Comprehension', in W. G. Chase (ed.), *Visual Information Processing*, New York: Academic Press, 1973.
Bridgemen, B., 'Visual Receptive Fields Sensitive to Absolute and Relative Motion During Tracking', *Science* 178 (1972), 1106-1108.
Brody, N. and Oppenheim, P., 'Tensions in Psychology Between the Methods of Behaviorism and Phenomenology', *Psycholog. Rev.* 73 (1966), 295-305.
Broglie, L., *The Current Interpretation of Wave Mechanics: A Critical Study*, Amsterdam: Elsevier, 1964.
Brown, D. G., 'Misconceptions of Inference', *Analysis* 15 (1955), 135-144.
Brown, R., *Words and Things*, Glencoe, Ill.: The Free Press, 1958.

Bruter, C. P., *Topologie et Perception*, Paris: Doin-Maloine S. A., 1974.
Buchler, J., *The Philosophy of Peirce: Selected Papers*, 1940.
Bugelski, B. R., 'Images as Mediators in One-Trial Paired-Associate Learning. II: Self-Timing in Successive Lists,' *J. Exp. Psychol.* 77 (1968), 328–334.
Bugelski, B. R., 'Words and Things and Images', *Am. Psychol.* 25 (1970), 1002–1012.
Bugelski, B. R., 'The Image as Mediator in One-Trial Paired-Associate Learning. III. Sequential Functions in Serial Lists,' *J. Exp. Psychol.* 103 (1974), 298–303.
Burke, L., 'On the Tunnel Effect,' *Quart. J. Exp. Psychol.* 4 (1952), 121–138.
Campbell, F. W., Cooper, G. F., and Enroth-Cugell, C., 'The Spatial Selectivity of the Visual Cells of the Cat,' *J. Physiol.* 203 (1969), 223–235.
Campbell, F. W., Cooper, G. F., Robson, J. G., and Sachs, M. B., 'The Spatial Selectivity of Visual Cells of the Cat and the Squirrel Monkey,' *J. Physiol.* 204 (1969b), 120–121.
Campbell, F. W. and Kulikowski, J. J., 'Orientational Selectivity of the Human Visual System,' *J. Physiol.* 187 (1966), 437–445.
Campbell, F. W. and Robson, J. G., 'Application of Fourier Analysis to the Visibility of Gratings,' *J. Physiol.*, 197 (1968), 551–566.
Capek, M., 'The Main Difficulties of the Identity Theory,' *Scientia* 36 (1969), 388–404.
Chang, H. T., Ruch, T. C., and Ward, A. A., Jr., 'Topographical Representation of Muscles in Motor Cortex in Monkeys,' *J. Neurophysiol.* 10 (1947), 39–56.
Chase, W. G. and Clarke, H. H., 'Mental Operations in the Comparison of Sentences and Pictures,' in L. Gregg (ed.), *Cognition in Learning and Memory*, New York: Wiley, 1972.
Chomsky, N., *Syntactic Structures*, Mouton, The Hague, 1957.
Chomsky, N., *Current Issues in Linguistic Theory*, The Hague: Mouton, 1964.
Chomsky, N., *Aspects of the Theory of Syntax*, Cambridge: M.I.T. Press, 1965.
Clark, H. H., 'Linguistic Processes in Deductive Reasoning,' *Psycholog. Rev.* 76 (1969), 387–404.
Clark, R., 'Sensuous Judgments,' *Noûs* 7 (1973), 45–56.
Clowes, M. B., 'On Seeing Things,' *Artificial Intelligence* 2 (1971), 79.
Collins, A. M. and Quillian, M. R., 'How to Make a Language User,' in E. Tulving and W. Donaldson (eds.), *Organization of Memory*, New York: Academic, 1972.
Cooper, L. A., 'Mental Rotation of Random Two-Dimensional Shapes,' *Cognitive Psychol.* 7 (1975), 20–43.
Cooper, L. A. and Shepard, R. N., 'Chronometric Studies of the Rotation of Mental Images,' in W. G. Chase (ed.), *Visual Information Processing*, New York: Academic, 1973. pp. 75–176.
Cornman, J., 'On the Elimination of "Sensations" and Sensations,' *Rev. Metaphys.* 22 (1968), 15–35.
Danto, A. C., 'Representational Properties and Mind-Body Identity,' *Rev. Metaphys.* 26 (1973), 401–411.
Davies, J. M. and Isard, S. D., 'Utterances as Programs,' in B. Meltzer and D. Michie (eds.), *Machine Intelligence* 7, Edinburgh: University Press, 1972.
Dewson, J. H. III, 'Cortical Responses to Patterns of Two-Point Cutaneous Stimulation,' *J. Comp. Physiol. Psychol.* 58 (1964), 387–389.
Eccles, J. C., 'Brain and Mind,' *Social Res.* 39 (1972), 753–757.

Enroth-Cugell, D. and Robson, J. G., 'The Contrast Sensitivity of Retinal Ganglion Cells of the Cat,' *J. Physiol.* **187** (1966), 517–552.

Evarts, E. V., 'Representation of Movements and Muscles by Pyramidal Tract Neurons of the Precentral Motor Cortex,' in M. D. Yahr and D. P. Purpura (eds.), *Neurophysiological Basis of Normal and Abnormal Motor Activities*, New York: Raven Press, 1967, pp. 215–254.

Evarts, E. V., 'Relation of Pyramidal Tract Activity to Force Exerted during Voluntary Movement,' *J. Neurophysiol.* **31** (1968), 14–27.

Feigenbaum, E. A. and Feldman, J. (eds.), *Computers and Thought*, New York: McGraw-Hill, 1963.

Fiegl, H., 'Mind-Body, not a Pseudo Problem,' in S. Hook (ed.), *Dimensions of Mind: A Symposium*, New York: New York University Press, 1960.

Feigl, H., *The 'Mental' and the 'Physical': The Essay and a Postscript*, Minneapolis: University of Minnesota Press, 1967.

Feigl, H., 'Some Crucial Issues of Mind-Body Monism,' *Synthèse* **22** (1971), 295–312.

Feyerabend, P. K. and Maxwell, G. (eds.), *Mind, Matter, and Method*, Minneapolis: University of Minnesota Press, 1966.

Feynman, R. P., Leighton, R. B., and Sands, M. (eds.), *The Feynman Lectures on Physics. Quantum Mechanics, Vol. III*, Reading, Massachusetts: Addison-Wesley Publishing Co., 1965.

Flavell, J. H., *The Developmental Psychology of Jean Piaget*, New York: Van Nostrand, 1963.

Fodor, J. A., (Review of *Symbol Formation* by H. Werner and B. Kaplan), *Language* **40** (1964), 566–578.

Fodor, J. A., Bever, T. G., and Garrett, M., *The Psychology of Language*, New York: McGraw-Hill, in press.

Foster, C. and Swanson, J. W. (eds.), *Experience and Theory*, Amherst: 1970.

Franks, J. J. and Bransford, J. D., 'The Acquisition of Abstract Ideas,' *Journal of Verbal Learning and Verbal Behavior* **11** (1972), 311–315.

Franks, J. J. and Bransford, J. D., 'A Brief Note on Linguistic Information,' *Journal of Verbal Learning and Verbal Behavior* **13** (1974), 217–219.

Frege, G., 'Begriffsschrift', in P. Geach and M. Black (eds.), *Translations from the Philosophical Writings of Gottlob Frege*, Oxford: Blackwell, 1960.

Freides, D. and Hayden, S. D., 'Monocular Testing: A Methodological Note on Eidetic Imagery,' *Perceptual and Motor Skills* **23** (1966), 88.

Frijda, N. H., 'Simulation of Human Long-Term Memory,' *Psycholog. Rev.* **77** (1972), 1–31.

Furth, H. G., *Thinking Without Language*, New York: Free Press, 1966.

Gaffron, M., 'Some New Dimensions in the Phenomenal Analysis of Visual Experience,' *J. Personality* **24** (1956), 285–307.

Gallie, W. B., *Peirce and Pragmatism*, 1952.

Garner, W. R., *Uncertainty and Structure as Psychological Concepts*, New York: John Wiley, 1962.

Gazzaniga, M. S., *The Bisected Brain*, New York: Appleton-Century-Crofts, 1970.

Gelernter, H., 'Realization of a Geometry-Theorem Proving Machine,' in E. A. Feigenbaum and J. Feldman (eds.), *Computers and Thought*, New York: McGraw-Hill, 1963.

Gellerman, L. W., 'From Discrimination in Chimpanzees and Two-Year-Old Children: I. Form (Triangularity) *per se*,' *J. Genetic Psychol.* **42** (1933), 3–27.
Gerbrandt, L. K., Spinelli, D. N., and Pribram, K. H., 'The Interaction of Visual Attention and Temperal Cortex Stimulation on Electrical Activity Evoked in the Striate Cortex,' *Electroenceph. Clin. Neurophysiol.* **29** (1970), 146–155.
Gesteland, R. C., Lettvin, J. Y., Pitts, W. H., and Chung, S. H., 'A Code in the Nose,' in H. L. Oestreicher and D. R. Moore (eds.), *Cybernetic Problems in Bionics*, New York: Gordon and Breach, 1968, pp. 313–322.
Gibson, E. J., *Principles of Perceptual Learning and Development*, New York: Appleton-Century-Crofts, 1969.
Gibson, J. J., *The Senses Considered as Perceptual Systems*, Boston: Houghton-Mifflin, 1966.
Gibson, J. J., 'New Reasons for Realism,' *Synthèse* **17** (1967), 162–172.
Gibson, J. J., 'What Gives Rise to the Perception of Motion?', *Psycholog. Rev.* **75** (1968), 335–346.
Gibson, J. J., Kaplan, G. A., Reynolds, H. N., and Wheeler, K., 'The Change from Visible to Invisible: A Study of Optical Transitions,' *Percept. Psychophys.* **5** (1969), 113–116.
Glanzer, M. and Peters, S. C., 'Reexamination of the Serial Position Affect,' *J. Exp. Psychol.* **64** (1962), 258–266.
Glezer, V. D., Ivanoff, V. A., and Tscherbach, T. A., 'Investigation of Complex and Hypercomplex Receptive Fields of Visual Cortex of the Cat as Spatial Frequency Filters,' *Vision Res.* **13** (1973), 1875–1904.
Globus, G. G., 'Consciousness and Brain. I. The Identity Thesis,' *Archives of General Psychiatry* **29** (1973), 153–160.
Good, I. J., 'Speculations Concerning the First Ultra-Intelligent Machine,' in F. L. Alt and M. Rubinoff (eds.), *Advances in Computers*, Vol. 6, New York: Academic Press, 1965.
Goodman, N., *Languages of Art*, Oxford: Oxford University Press, 1969.
Goudge, T. A., *The Thought of C.S. Peirce*, 1950.
Granit, Ragnar, *The Basis of Motor Control*, New York: Academic Press, 1970.
Green, C. and Raphael, R., 'The Use of Theorem Proving Techniques in Question Answering Systems,' in *Proceedings of the National Conference*, New York: Association for Computing Machinery, 1968.
Gregory, R. L., *Eye and Brain*, New York: McGraw-Hill Book Co., 1966.
Groot, A. D. de, 'Perception and Memory Versus Thought: Some Old Ideas and Recent Findings,' in B. Kleinmuntz (eds.), *Problem Solving*, New York: Wiley, 1966.
Groot, A. D. de, *Methodology*, The Hague: Mouton, 1969.
Gross, C. G., 'Visual Functions of Inferotemporal Cortex,' in R. Jung (ed.), *Handbook of Sensory Physiology*, 7 Part 3B, Berlin: Springer-Verlag, 1972.
Guilford, J. P., *The Nature of Human Intelligence*, New York: McGraw-Hill, 1967.
Guilford, J. P., Fruchter, B., and Zimmerman, W. S., 'Factor Analysis of the Army Air Force Sheppard Field Battery of Experimental Aptitude Tests,' *Psychometrika* **17** (1952), 45–68.
Gunderson, K., 'Robots, Consciousness, and Programmed Behavior,' *British J. Phil. Sci.* **19** (1968), 109–122..

Haber, R. N. and Standing, L., 'Clarity and Recognition of Masked and Degrading Stimuli,' *Psychonomic Sci.* **13** (1968), 83–84.
Haber, R. N., 'Where Are the Visions in Visual Perception?', in S. J. Segal (ed.), *Imagery: Current Cognitive Approaches*, New York: Academic Press, 1971.
Haber, R. N. and Haber, R. B., 'Eidetic Imagery: I. Frequency,' *Perceptual and Motor Skills* **19** (1964), 131–138.
Hall, E. W., 'The Adequacy of a Neurological Theory of Perception,' *Phil. Phenomen. Res.* **20** (1959–1960), 75–84.
Hall, E. W., *Our Knowledge of Fact and Value*, Chapel Hill: University of North Carolina Press, 1961.
Hanson, N. R., *Patterns of Discovery*, Cambridge: University of Cambridge Press, 1958.
Hanson, N. R., 'On Having the Same Visual Experiences,' *Mind* **69** (1960), 340–350.
Hayes, P., 'Robotologic,' *Machine Intelligence 5*, Meltzer, B. and Michie, D. (eds.), Edinburgh, 1969.
Hayek, F. A., *The Sensory Order: An Inquiry Into the Foundations of Theoretical Psychology*, London: Routledge & Kegan Paul, 1952.
Hebb, D. O., 'Emotion in Man and Animal: an Analysis of the Intuitive Processes of Recognition,' *Psycholog. Rev.* **53** (1946), 88–106.
Hebb, D. O., *The Organization of Behavior; A Neuro-Psychological Theory*, New York: Wiley, 1949.
Hebb, D. O., 'The American Revolution,' *Am. Psychol.* **15** (1960), 735–745.
Hebb, D. O., 'The Semiautonomous Process: Its Nature and Nurture,' *Am. Psychol.* **18** (1963), 16–27.
Hebb, D. O., *A Textbook of Psychology*, (2nd. ed.), Philadelphia: Saunders, 1966.
Hebb, D. O., 'Concerning Imagery,' *Psycholog. Rev.* **75** (1968), 466–477.
Hebb, D. O., 'The Mind's Eye,' *Psychol. Today* **2** (1969), 54–57, 67–68.
Hebb, D. O., *A Textbook of Psychology*, (3rd ed.), Philadelphia: Saunders, 1972.
Heiseberg, W., *Physics and Philosophy*, London: G. Allen and Unwin, 1959.
Hewitt, C., *Description and Theoretical Analysis (using schemata) of PLANNER*, Unpublished doctoral dissertation, Massachusetts Institute of Technology, Cambridge, Massachusetts, 1971.
Hirsch, H. and Spinelli, D. N., 'Distribution of Receptive Field Orientation: Modification Contingent on Conditions of Visual Experience,' *Science* **168** (1970), 869–871.
Hochberg, J., 'Attention, Organization, and Consciousness,' in D. I. Mostofsky (ed.), *Attention: Contemporary Theory and Analysis*, New York: Appleton-Century-Crofts, 1970.
Holt, R. R., 'Imagery: The Return of the Ostracized,' *Am. Psychol.* **19** (1964), 254–264.
Horn, G., Stechler, G., and Hill, R. M., 'Receptive Fields of Units in the Visual Cortex of the Cat in the Presence and Absence of Bodily Tilt,' *Exp. Brain Res.* **15** (1972), 113–132.
Horowitz, M. J., *Image Formation and Cognition*, New York: Appleton-Century-Crofts, 1970.
Hubel, D. H. and Wiesel, T. N., 'Receptive Fields and Functional Architecture of Monkey Striate Cortex,' *J. Physiol.* **195** (1968), 215–243.
Humphrey, G., *Thinking: An Introduction to Its Experimental Psychology*, London: Methuen, 1951.

BIBLIOGRAPHY

Humphrey, N. K. and Weiskrantz, L., 'Size Constancy in Monkeys with Inferotemporal Lesions,' *Quart. J. Exp. Psychol.* **21** (1969), 225–238.
James, W., *Principles of Psychology*, New York: Dover, 1950.
Johansson, G., 'Visual Perception of Biological Motion and a Model for its Analysis,' *Percept. Psychophys.* **14** (1973), 201–211.
Kant, I., *Critique of Pure Reason*, transl. by N. K. Smith, London: Macmillan, 1958.
Katz, J. J. and Fodor, J. A., 'The Structure of a Semantic Theory,' *Language* **39** (1963), 170-210.
Kintsch, W. and Monk, D., 'Storage of Complex Information in Memory: Some Implications of the Speed with Which Inferences Can be Made,' *J. Exp. Psychol.* **94** (1972), 25–32.
Koffka, K., *Principles of Gestalt Psychology*, New York: Harcourt, Brace, 1935.
Köhler, W., 'Psychological Remarks on Some Questions of Anthropology,' *Am. J. Psychol.* **50** (1937), 271–288.
Köhler, W., *Gestalt Psychology*, New York: Liveright, 1929, (revised edition, 1947).
Köhler, W., *The Place of Value in a World of Facts*, New York: Liveright, 1938.
Köhler, W., 'The Present Situation in Brain Physiology,' *Am. Psychol.* **13** (1958), 150–154.
Köhler, W., 'A Task for Philosophers,' in Feyerabend and Maxwell, 1966.
Lakatos, I., 'Proofs and Refutations,' *Brit. J. Phil. Sci.* **14** (1963–4), 1–25.
Lambert, W. E. and Rawlings, C., 'Bilingual Processing of Mixed-Language Associative Networks,' *Journal of Verbal Learning and Verbal Behavior* **8** (1969), 604–609.
Lashley, K. S., *Brain Mechanisms and Intelligence*, Chicago: University of Chicago Press, 1929.
Lashley, K. S., 'In Search of the Engram,' in *Symposia of the Society of Experimental Biology*, No. 4, Cambridge: Cambridge University Press, 1950.
Lashley, K. S., 'Persistent Problems in the Evolution of Mind,' in F. A. Beach, H. W. Nissen and E. G. Boring (eds.), *The Neuropsychology of Lashley*, New York: McGraw-Hill Book Co., 1960, pp. 455–477.
Lawden, D. F., 'Towards a Non-Behavioral Psychology,' *Philosophic. J.* **9** (1972), 116–129.
Lewis, D. K., 'An Argument for the Identity Theory,' *J. Philos.* **69** (1966), 17–25.
Leask, J., Haber, R. N., and Haber, R. B., 'Eidetic Imagery in Children: II. Longitudinal and Experimental Results,' *Psychonomic Monograph Supplements* **3** (1969), 25–48.
Lewis, D., 'Psychophysical and Theoretical Identifications,' *Australasian Journal of Philosophy* **50** (1972), 249–258.
Lindsay, R. K., 'Inferential Memory as the Basis of Machines which Understand Natural Language,' in Feigenbaum and Feldman, 1963.
Liss, P., 'Does Backward Masking by Visual Noise Stop Stimulus Processing?', *Percept. Psychophys.* **4** (1968), 328–330.
Locke, D., 'Must a Materialist Pretend he's Anesthetized?', *Philosophic. Quart.* **21** (1971), 217–231.
Lynch, J. C., *A Single Unit Analysis of Contour Enhancement in the Somesthetic System of the Cat*, Ph. D. Thesis, Stanford University, Neurological Sciences, 1971.
McCarthy, J., 'Programs with Common Sense,' *Proceedings of the Symposium on Mechanization of Thought Processes*, London: Her Majesty's Stationery Office, 1959.

McCarthy, J. and Hayes, P. 'Some Philosophical Problems from the Standpoint of Artificial Intelligence,' *Machine Intelligence* 4, B. Meltzer and D. Michie (eds.), Edinburgh, 1969.
MacNamara, J., 'Cognitive Basis of Language Learning in Infants,' *Psycholog. Rev.* 79 (1972), 1–13.
Maffei, L. and Fiorentini, A., 'The Visual Cortex as a Spatial Frequency Analyzer,' *Vision Res.* 13 (1973), 1255–1267.
Malmo, R. B. and Surwillo, W. W., 'Sleep Deprivation: Changes in Performance and Physiological Indicants of Activation,' *Psychological Monographs* 74 (1960), 502.
Maxwell, G., 'Structural Realism and the Meaning of Theoretical Terms,' in M. Radner and S. Winokur (eds.), *Minnesota Studies in the Philosophy of Science*, Vol. 4, Minneapolis: University of Minnesota Press, 1970a.
Maxwell, G., 'Theories, Perception, and Structural Realism,' in R. G. Coldny (ed.), *The Nature and Function of Scientific Theories*, Pittsburgh: University of Pittsburgh Press, 1970b.
Maxwell, G., 'Russell on Perception: A Study of Philosophical Method,' in D. F. Pears (ed.), *Bertrand Russell: A Collection of Critical Essays*, Garden City, New York: Doubleday, 1972.
Meehl, P. E., 'The Compleat Autocerboscopist: A Thought-Experiment on Professor Feigl's Mind-Body Identity Thesis,' in P. K. Feyerabent and G. Maxwell (eds.), *Mind, Matter, and Method: Essays in Philsophy and Science In Honor of Herbert Feigl*, Minneapolis: University of Minnesota Press, 1966.
Michotte, A., Thinès, G., and Crabbé, G., *Les Complements Amodaux des Structures Perceptives*, Louvain, Belgium: Publications Université Louvain, 1964.
Miller, G. A., Galanter, E. H., and Pribram, K. Y., *Plans and the Structure of Behavior* New York: Henry Holt and Co., 1960.
Milner, B., 'Psychological Defects Produced by Temporal Lobe Excision,' in H. C. Solomon, S. Cobb, and W. Penfield (eds.), *The Brain and Human Behavior* 36 (1958), 244–257. (Proceedings of the Association for Research in Nervous and Mental Disease), Baltimore: The Williams & Wilkins Co.
Milner, B., 'Hemisphere Specialization: Scope and Limits.' in F. O. Schmitt and F. G. Worder (eds.), *The Neurosciences, Third Study Program*, Cambridge, Massachusetts: The MIT Press, 1974, pp. 75–89.
Minsky, M. L., 'Steps Towards Artificial Intelligence,' in Feigenbaum and Feldman. 1963.
Minsky, M. L. (ed.), *Semantic Information Processing*, Cambridge, Mass.: M.I.T. Press, 1968.
Minsky, M. L., 'Descriptive Languages and Problem Solving,' in Minsky, 1968.
Minsky, M. L., *Introduction* to Minsky, 1968.
Minsky, M. L., 'Form and Content in Computer Science,' ACM Turing Lecture, *J. A. C. M.* 17 (1970), 197–215.
Minsky, M. L. and Papert, S., *Perceptions: An Introduction to Computational Geometry*, Cambridge: The MIT Press, 1969.
Mishkin, M., 'Cortical Visual Areas and Their Interaction.' in A. G. Karczmar and J. C. Eccles (eds.), *The Brain and Human Behavior,* Berlin Springer-Verlag, 1972.
Moran, T. P., *The Symbolic Imagery Hypothesis: A Production System Model*, Unpublished doctoral dissertation, Carnegie-Mellon University, 1973a.

Moran, T. P., 'The Symbolic Nature of Visual Imagery,' in *Third International Joint Conference on Artifical Intelligence*, Menlo, Park, California: Stanford Research Institute, 1973b, pp. 472-477.
Morris, G. O., Williams, H. L., and Lubin, A., 'Misperception and Disorientation During Sleep Deprivation,' *A. M. A. Archives of General Psychiatry* 2 (1960), 247-254.
Moseley, A. L., 'Hypnagogic Hallucinations in Relation to Accidents.' *Am. Psychol.* 8 (1953), 407.
Moyer, R. S., *On the Possibility of Localizing Visual Memory*, Ph. D. Thesis, Stanford University, 1970.
Moyer, R. S., 'Comparing Objects in Memory: Evidence Suggesting an Internal Psychophysics,' *Percept. Psychophys.* 13 (1973), 180-184.
Mueller, I., 'Euclid's Elements and the Axiomatic Method,' *Brit. J. Phil. Sci.* 20 (1969), 289-309.
Nagel, T., 'Physicalism,' *Philosophic. Rev.* 74 (1965), 339-356.
Natsoulas, T., 'Concerning Introspective "knowledge",' *Psychol. Bull.* 73 (1970), 89-111.
Neisser, U., *Cognitive Psychology*, New York: Appleton-Century-Crofts, 1967.
Neisser, U., 'Changing Conceptions of Imagery,' in P. W. Sheehan (ed.), *The Function and Nature of Imagery*, New York: Academic Press, 1972.
Neisser, U. and Kerr, N., 'Spatial and Mnemonic Properties of Visual Images,' *Cognitive Psychol.* 5 (1973), 138-150.
Newell, A., 'On the Analysis of Human Problem Solving Protocols,' in J. C. Gardin and B. Jaulin (eds.), *Calcul et Formalisation dans les Sciences de l'Homme*, Paris: CNRS, 1968, pp. 145-185.
Newell, A. and Simon, H., *Human Problem Solving*, Englewood Cliffs, N. J.: Prentice-Hall, 1972.
Newell, A., 'A Theoretical Exploration of Mechanisms for Coding The Stimulus,' in A. W. Melton and E. Martin (eds.), *Coding Processes in Human Memory*, Washington, D.C.: Winston, 1972, pp. 373-434.
Newell, A., 'Artificial Intelligence and the Concept of Mind,' in R. C. Schank and K. M. Colby (eds.), *Computer Models of Thought and Language*, San Francisco: Freeman, 1973a, pp. 1-60.
Newell, A., 'Production Systems: Models of Control Structures,' in W. G. Chase (ed.), *Visual Information Processing*, New York: Academic, 1973b, pp. 463-526.
Neumann, J. von, *Mathematische Grundlagen der Quantenmechanik*, Berlin: Springer-Verlag, 1932.
Norman, D. A., 'Toward a Theory of Memory and Attention,' *Psychol. Rev.* 75 (1968), 522-536.
O'Neil, W. M., 'Basic Issues in Perceptual Theory,' *Psychol. Rev.* 65 (1958), 348-361.
Ong, W. J., *Ramus, Method, and the Decay of Dialogue*, Cambridge: Harvard University Press, 1958.
Osgood, C. E., Suci, G. J., and Tannenbaum, P. H., *The Measurement of Meaning*, Urbana, Ill.: University of Illinois Press, 1957.
Paivio, A. U., 'Mental Imagery in Associative Learning and Memory,' *Psychol. Rev.* 76 (1969), 241-263.
Paivio, A. U., *Imagery and Verbal Processes*, New York: Holt, Rinehart and Winston, 1971.

Penfield, W. and Boldrey, E., 'Somatic Motor and Sensory Representation in the Cerebral Cortex of Man as Studied by Electrical Stimulation,' *Brain* **60** (1937), 389–443.
Pepper, S. C., 'A Neural-Identity Theory of Mind,' in S. Hook (ed.), *Dimensions of Mind: A Symposium*, New York: New York University Press, 1960.
Perkins, M., Matter, Sensation, and Understanding,' *Am. Philosophic. Quart.* **8** (1971), 1–12.
Pfaffmann, C. C., 'The Pleasures of Sensation,' *Psychol. Rev.* **67** (1960), 253–268.
Phillips, C. G., 'Changing Concepts of the Precentral Motor Area,' in J. C. Eccles (ed.), *Brain and Conscious Experience*, New York: Springer-Verlag 1965, 389–421.
Pollen, D. A., Lee, J. R., and Taylor, J. H., 'How Does the Striate Cortex Begin the Reconstruction of the Visual World?', *Science* **173** (1971), 74–77.
Pollen, D. A. and Taylor, J. H., 'The Striate Cortex and the Spatial Analysis of Visual Space,' in *The Neurosciences Study Program, Vol. III*, Cambridge, Massachusetts: The MIT Press, 1974, pp. 239–247.
Poppen, R., Pribram, K. H. and Robinson, R. S., 'The Effects of Frontal Lobotomy in Man on Performance of a Multiple Choice Task,' *Exp. Neurol.* **11** (1965), 217–229.
Pribram, H. and Barry, J., 'Further Behavioral Analysis of the Parieto-Temporo-Preoccipital cortex,' *J. Neurophysiol.* **19** (1956), 99–106.
Pribram, K. H., 'Toward a Science of Neuropsychology: (Method and Data),' in R. A. Patton (ed.), *Current Trends in Psychology and the Behavioral Sciences*, Pittsburgh: University of Pittsburgh Press, 1954, pp. 115–142.
Pribram, K. H., 'Neocortical Function in Behavior,' in H. F. Harlow and C. N. Woolsey (eds.), *Biological and Biochemical Bases of Behavior*, Madison: University of Wisconsin Press, 1958, pp. 151–172.
Pribram, K. H., 'A Review of Theory in Physiological Psychology,' in *Ann. Rev. Psychol.* **2** (1960), 1–40, Palo Alto: Annual Reviews, Inc.
Pribram, K. H., 'Proposal for a Structural Pragmatism: Some Neuropsychological Considerations of Problems in Philosophy,' in B. Wolman and E. Nagel (eds.), *Scientific Psychology: Principles and Approaches*, New York: Basic Books, 1965, pp. 426–459.
Pribram, K. H., 'The Amnestic Syndromes: Disturbances in Coding?', in G. A. Talland and N. C. Waugh (eds.), *Pathology of Memory*, New York: Academic Press, pp. 127–157, 1969.
Pribram, K. H., *Languages of the Brain: Experimental Paradoxes and Principles in Neuropsychology*, Englewood Cliffs, N.J.: Prentice-Hlll, Inc., 1971a.
Pribram, K. H., 'The Realization of Mind,' *Synthese* **22** (1971b), 313–322.
Pribram, K. H., 'How Is It That Sensing So Much We Can Do So Little?', in *The Neurosciences Study Program, III*, Cambridge, Massachusetts: The MIT Press, 1974a, pp. 249–261.
Pribram, K. H., 'The Isocortex,' in D. A. Hamburg and H. K. H. Brodie (eds.), *American Handbook of Psychiatry, Volume 6*, New York: Basic Books, 1974b.
Pribram, K. H., 'Problems Concerning the Structure of Consciousness,' in G. Globus, G. Maxwell and I. Savodnik (eds.), *Scientific and Philosophical Approaches to the Structure of Consciousness and Brain*, New York: Plenum Press (in press).

Pribram, K. H., Kruger, L., Robinson, R., and Berman, A. J., 'The Effects of Precentral Lesions on the Behavior of Monkeys,' *Yale J. Biol. Med.* **28** (1955–56), 428–443.
Pribram, K. H., Nuwer, M., and Baron, R., 'The Holographic Hypothesis of Memory Structure in Brain Function and Perception,' in R. C. Atkinson, D. H. Krantz, R. C. Luce, and P. Suppes (eds.), *Contemporary Developments in Mathematical Psychology*, San Francisco: W. H. Freeman and Co., 1974.
Pribram, K. H., Spinelli, D. N., and Kamback, M. C., 'Electrocortical Correlates of Stimulus Response and Reinforcement,' *Science* **157** (1967), 94–96.
Pribram, K. H., Spinelli, D. N., and Reitz, S. L., 'Effects of Radical Disconnextion of Occipital and Temporal Cortex on Visual Behavior of Monkeys,' *Brain* **92** (1969), 301–312.
Pritchard, R. M., 'Stabilized Images on the Retina,' *Scient. Am.* **204** (1961), 72–78.
Pritchard, R. M., Heron, W., and Hebb, D. O., 'Visual Perception Approached by the Method of Stabilized Images,' *Canad. J. Psychol.* **14** (1960), 67–77.
Pylyshyn, Z. W., *Temporal Factors in Immedaite Memory*, Unpublished doctoral dissertation, University of Saskatchewan, Saskatoon, Canada, 1963.
Pylyshyn, Z. W., 'Competence and Psychological Reality,' *Am. Psychol.* **27** (1972), 546–552.
Pylyshyn, Z. W., 'The Role of Competence Theories in Cognitive Psychology,' *J. Psycholinguistic Res.* **2** (1973), 21–50.
Pylyshyn, Z. W., 'What the Mind's Eye Tells the Mind's Brain: A Critique of Mental Imagery,' *Psycholog. Bull.* **80** (1973), 1–24.
Quinton, A., 'Ryle on Perception,' in O. P. Wood and G. Pitcher (eds.), *Ryle: A Collection of Critical Essays*, Garden City, N.Y.: Doubleday, 1970.
Quinton, A., *The Nature of Things*, London: Routledge and Kegan Paul, 1973.
Racine, B., *La Transformation de l'Image Mentale*, Unpublished Master's Thesis, Université de Montréal, 1971.
Raphael, B., 'Programming a Robot,' in *Proceedings of the 1968 International Federation for Information Processing Congress*, Amsterdam: North Holland Publishing, 1968.
Raphael, B., 'The Frame Problem in Problem-Solving Systems,' in N. V. Findler and B. Meltzer (eds.), *Artificial Intelligence and Heuristic Programming*, Edinburgh: Edinburgh University Press, 1971.
Raphael, B., 'A Computer Program Which "Understands",' in Minsky, 1968.
Ratliff, F., *Mach Bands: Quantitative Studies in Neural Networks in the Retina*, San Francisco: Holden–Day, Inc., 1965.
Reese, H. W. (Chm.), 'Imagery in Children's Learning: A Symposium', *Psycholog. Bull.* **73** (1970), 404–414.
Reichenbach, H., *Experience and Prediction*, Chicago: University of Chicago Press, 1938.
Reitman, W., *Cognition and Thought*, New York: Wiley, 1965.
Reitman, J. S. and Bower, G. H., 'Storage and Later Recognition of Exemplars and Concepts,' *Cognitive Psychol.* **4** (1973), 194–206.
Reitz, S. L. and Pribram, K. H., 'Some Subcortical Connections of the Inferotemporal Gyrus of Monkey,' *Exp. Neurol.* **25** (1969), 632–645.
Richardson, A., *Mental Imagery*, New York: Springer, 1969.

Reynolds, H. N., 'Temporal Estimation in the Perception of Occluded Motion,' *Perceptual and Motor Skills* **26** (1968), 407–416.
Roberts, L. G., 'Machines Perception of Three-Dimensional Solids,' in J. T. Tipett *et al.* (eds.), *Optical and Electro-Optical Information Processing*, Cambridge, Mass.: M.I.T. Press, 1965.
Rorty, R., 'Mind-Body Identity, Privacy, and Categories,' *Rev. Metaphys.* **19** (1965), 24–54.
Rorty, R., 'In Defense of Eliminative Materialism,' *Rev. Metaphys.* **24** (1970), 112–121.
Rosch, E., 'Universals and Cultural Specifics in Human Categorization,' in W. S. Lonner and R. Breslin (eds.), *Cross-Cultural Perspectives on Learning*, London: Sage Publications, 1974.
Rosenberg, S., 'Modeling Semantic Memory: Effects of Presenting Semantic Information In Different Modalities,' Doctoral thesis, Carnegie-Mellon University, 1974.
Rosenblueth, A., *Mind and Brain: A Philosophy of Science*, Cambridge, Mass.: MIT Press, 1970.
Rothblat, L. and Pribram, K. H., 'Selective Attention: Input Filter or Response Selection?', *Brain Res.* **39** (1972), 427–436.
Rozeboom, W. W., 'Formal Analysis and the Structure of Behavior Theory,' in H. Feigl and G. Maxwell (eds.), *Current Issues in the Philosophy of Science; Symposia of Scientists and Philosophers*, New York: Holt, Rinehart and Winston, 1961.
Rozeboom, W. W., 'Problems in the Psycho-Philosophy of Knowing,' in J. R. Royce and W. W. Rozeboom (eds.), *The Psychology of Knowing*, New York: Gordon and Breach, 1972.
Rumelhart, D. E., Lindsey, P. H., and Norman, D. A., 'A Process Model for Long-Term Memory,' in E. Tulving and W. Donaldson (eds.), *Organization of Memory*, New York: Academic Press, 1972.
Russell, B., *An Inquiry Into Meaning and Truth*, New York: Norton, 1940.
Russell, B., 'Reply to Criticisms,' in P. Schilpp (ed.), *The Philosophy of Bertrand Russell*, Evanston, Ill.: Library of Living Philosophers, 1946.
Russell, B., *My Philosophical Development*, New York: Simon and Schuster, 1959.
Ryle, G., *The Concept of Mind*, New York: Barnes and Noble, 1949.
Saltz, E., *The Cognitive Bases of Human Learning*, Homewood, Ill.: Dorsey, 1971.
Schank, R. C., 'Conceptual Dependency: A Theory of Natural Language Understanding.' *Cognitive Psychol.* **3** (1972), 552–631.
Schlick, M., 'Form and Content: An Introduction to Philosophical Thinking,' *Gesammelte Aufsätze, 1926–1936*, Hildesheim, Germany: Verlag, 1969.
Schroedinger, E., 'Discussion of Probability Relations Between Separated Systems,' *Pro. Cambridge Philosophic. Soc.* **31** (1935), 555–563.
Segal, S. J. (ed.), *Imagery*, New York: Academic Press, 1971.
Sellars, R. W., 'Positivism and Materialism,' *Phil. Phenomenolog. Res.* **7** (1946), 12–41.
Sellars, W., 'Philosophy and the Scientific Image of Man,' in R. G. Colodny (ed.), *Frontiers of Science and Philosophy*, Pittsburgh: University of Pittsburgh Press, 1962.
Sellars, W., *Science, Perception and Reality*, New York: Humanities Press, 1963.
Sellars, W., 'The Identity Approach to the Mind-Body Problem,' *Review Metaphys.* **18** (1965), 430–451.

Sellars, W., 'The Double-Knowledge Approach to the Mind-Body Problem,' *New Scholasticism* **45** (1971a), 269–289.
Sellars, W., 'Science, Sense Impressions, and Sensa: A Reply to Cornman,' *Rev. Metaphys.* **24** (1971b), 391–447.
Shallice, T., 'Dual Functions of Consciousness,' *Psycholog. Rev.* **79** (1972), 383–393.
Sheehan, P. W. (ed.), *The Function and Nature of Imagery*, New York: Academic Press, 1972.
Shepard, R. N., 'Learning and Recall as Organization and Search,' *Journal of Verbal Learning and Verbal Behavior* **5** (1966), 201–204.
Shepard, R. N., and Chipman, S., 'Second-Order Isomorphism of Internal Representations: Shapes of States,' *Cognitive Psychol.* **1** (1970), 1–17.
Sibley, F. (ed.), *Perception*, London: Methuen, 1971.
Siipola, E. M. and Hayden, S. D., 'Exploring Eidetic Imagery Among the Retarded,' *Perceptual and Motor Skills* **21** (1965), 275–286.
Simmel, M. L., 'Phantoms in Patients with Leprosy and in Elderly Digital Amputees,' *Am. J. Psychol.* **69** (1956), 529–545.
Singer, M. and Rosenberg, S. T., 'The Role of Grammatical Relations in the Abstraction Linguistic Ideas,' *J. Verbal Learning Verbal Behavior* **12** (1973), 237–284.
Skinner, B. F., 'Behaviorism at Fifty,' *Science* **140** (1963), 951–958.
Sloman, A., 'Explaining Logical Necessity,' *Proc. Aristotelian Soc.* **69** (1968–9), 33–50.
Sloman, A., 'Tarski, Grege and the Liar Paradox,' *Philosophy* **46** (1971), 133–147.
Smart, J. J. C., 'Sensations and Brain Processes,' *Philosophic. Rev.* **68** (1959), 141–156.
Smart, J. J. C., 'Reports of Immediate Experience,' *Synthese* **22** (1971), 346–359.
Smart, J. J. C., 'Further Thoughts on the Identity Theory,' *The Monist* **56** (1972), 149–162.
Sperling, G., 'A Model for Visual Memory Tasks,' *Human Factors* **5** (1963), 19–31.
Sperling, G., 'Successive Approximations to a Model for Short-Term Memory,' *Acta Psychologica* **27** (1967), 285–292.
Simon, H., *The Sciences of the Artificial*. Cambridge, Mass.: M.I.T. Press, 1969.
Simon, H. A. and Barenfeld, M., 'Information-Processing Analysis of Perceptual Processes in Problem Solving,' *Psycholog. Rev.* **76** (1969), 473–483.
Simon, H. A. and Newell, A., 'The Uses and Limitations of Models,' in L. D. White (ed.), *The State of the Social Sciences*, Chicago: University of Chicago Press, 1956.
Slagle, J. R., *Artificial Intelligence: The Heuristic Programming Approach*, New York: McGraw-Hill, 1971.
Sperry, R. W., 'Neurology and the Mind-Brain Problem,' *Am. Scientist* **40** (1952), 291–312.
Sperry, R. W., 'Mind Brain and Humanist Values,' in J. R. Platt (ed.), *New Views of the Nature of Man*, Chicago: University of Chicago Press, 1965.
Sperry, R. W., 'A Modified Concept of Consciousness,' *Psycholog. Rev.* **76** (1969), 532–536.
Sperry, R. W., 'An Objective Approach to Subjective Experience: Further Explanation of a Hypothesis,' *Psycholog. Rev.* **77** (1970), 585–590.
Sperry, R. W., 'Lateral Specialization in the Surgically Separated Hemispheres,' in F. O.

Schmitt and F. G. Worden (eds.), *The Neurosciences Third Study Program*, Cambridge, Massachusetts: The MIT Press, 1974, pp. 5–19.

Spinelli, D. N., 'Recognition of Visual Patterns,' in *Perception and Its Disorders*, Research publication of the Association for Research in Nervous and Mental Disease, Volume 48, 1970, pp. 139–149.

Spinelli, D. N. and Pribram, K. H., 'Changes in Visual Recovery Functions Produced by Temporal Lobe Stimulation in Monkeys,' *Electroenceph. Clin. Neurophysiol.* **20** (1966), 44–49.

Spinelli, D. N. and Pribram, K. H., 'Changes in Visual Recovery Function and Unit Activity Produced by Frontal Cortex Stimulation,' *Electroenceph. Clin. Neurophysiol.* **22** (1967), 143–149.

Staats, A. W., *Learning, Language, and Cognition*, New York: Holt, Rinehart and Winston, 1968.

Stout, G. F., *Mind and Matter*, New York: Macmillan, 1931.

Stromeyer, C. F., III, 'Eidetikers,' *Psychology Today* **4** (1970), 76–80.

Stromeyer, C. F., III, and Psotka, J., 'The Detailed Texture of Eidetic Images,' *Nature* **225** (1970), 346–349.

Stromeyer, C. F., III, and Klein, S., 'Spatial-Frequency Channels in Human Vision as Asymmetric (Edge) Mechanisms,' *Vision Res.* **14** (1974), 1409–1420.

Stromeyer, C. F., III, and Klein, S., 'The Detectability of Frequency Modulated Gratings: Evidence Against Narrow-Band Spatial Frequency Channels in Human Vision,' *Vision Res.* **15** (1975), 899–910.

Stromeyer, C. F., III, and Klein, S., 'Adaptation to Complex Gratings: On Inhibition Between Spatial Frequency Channels,' *Vision Res.* (Submitted).

Sutherland, I. W., *Sketchpad: A Man-Machine Graphical Communication System*, (American Federation of Information Processing Societies Spring Joint Computer Conference), Vol. 23. Baltimore, Md.: Spartan, 1963.

Taylor, C., 'The Opening Argument of the Phenomenology,' in A. MacIntyre (ed.), *Hegel: A Collection of Critical Essays*, Garden City, N.Y.: Doubleday, 1972.

Teuber, H. L., 'Perception,' in J. Field, H. W. Magoun, and V. E. Hall (eds.), *Handbook of Physiology: Neurophysiology*, Vol. 3, Washington, D.C.: American Physiological Society, 1960.

Thom, R., *Stabilité Structurelle et Morphogénèse*, Reading, Massachusetts: W. A. Benjamine, Inc., 1972.

Thornton, M. T., 'Ostensive Terms and Materialism,' *The Monist* **56** (1972), 193–214.

Tolman, E. C., 'Concerning the Sensation Quality: A Behavioristic Account,' *Psycholog. Rev.* **29** (1922), 140–145.

Tolman, E. C., *Purposive Behavior in Animals and Men*, Berkeley: University of California Press, 1932.

Tolman, E. C., 'Cognitive Maps in Animals and Man,' *Psycholog. Rev.* **55** (1948), 189–208.

Toulmin, S., *The Philosophy of Science*, London: Hutchinson, 1962.

Triesman, A. M., 'Verbal Cues, Language and Meaning in Selective Attention,' *Am. J. Psychol.* **77** (1964), 206–219.

Ungerleider, L., 'Deficits in Size Constancy Discrimination: Further Evidence for Dissociation between Monkeys with Inferotemporal and Prestriate Lesions,' Paper presented at the Eastern Psychological Association Convention, April, 1975.

Vesey, G., *Perception,* Garden City, N. Y.: Doubleday, 1971.
Watson, J. B., Psychology From the Standpoint of the Behaviourist, Philadelphia: Lippincott, 1919.
Weber, R. J. and Bach, M., 'Visual and Speech Imagery,' *British J. Psychol.* **60** (1969), 199–202.
Weber, R. J. and Harnish, R., 'Visual Imagery for Words: the Hebb Test,' *J. Exp. Psychol.* **102** (1974), 409–414.
Weimer, W. B., 'Psycholinguistics and Plato's Paradoxes of the Meno' *Am. Psychol.* **28** (1973), 15–33.
Weinberg, S., 'Unified Theories of Elementary-Particle Interaction,' *Scient. Am.* **231** (1974), 50–59.
Werblin, P. S. and Dowling, J. E., 'Organization of the Retina of the Mud Puppy, Necturus Maculsus, II' Intracellular Recording,' *J. Neurophysiol.* **32** (1969), 339–355.
Whitlock, D. G. and Nauta, W. J., 'Subcortical Projections from the Temporal Neocortex in Macaca-Mulatta,' *J. Comp. Neurol.* **106** (1956), 183–212.
Wilkerson, T. E., 'Seeing-As,' *Mind* **82** (1973), 481–496.
Wilson, M., 'Effects of Circumscribed Cortical Lesions upon Somesthetic and Visual Discrimination in the Monkey,' *J. Comp. Physiol. Psych.* **50** (1957), 630–635.
Winograd, T., 'Understanding Natural Language,' *Cognitive Psychol.* **3** (1972), 1–191.
Wittgenstein, L., *Philosophical Investigations*, Trans., G.E.M. Anscombe, Oxford: Blackwell, Ist. edn., 1953; 2nd. edn., 1958.
Wittgenstein, L., *Remarks on the Foundations of Mathematics,* Oxford: Blackwell, 1956.
Woodworth, R. S., *Experimental Psychology*, New York: Holt, 1938.
Yolton, J. W., 'The Form and Development of Experience,' *Acta Psychol.* **21** (1963), 357–370.

INDEX

Abstraction 5–7, 51–2, 146–8
Allport, G. W. 149
Alston, W. P. 251
Analysis-by-Synthesis 6
Anderson, J. R. 48, 53–5, 56, 65
Anderson, R. C. 52
Arnheim, R. 2
Association 5, 6, 37–46, 47, 139
Attention 6
Attribution 101–3
Atwood, G. 13–14
Ayer, A. J. 191

Bahrick, H. P. 14
Baron, R. 160
Baylor, George W. 31–4, 48, 73, 84
Bekesy, von, G. 179
Berkeley, Bishop 18, 146
Bernstein, N. 176–7
Bilingualism 42–6, 95, 97
Bindra, D. 234
Binet, A. 144, 152–3
Bohm, David 182–3
Boucher, B. 14
Bower, G. H. 2, 7, 98, 53–55, 58, 65
Bransford, J. D. 96
Brehm, J. 258
Brown, R. 48
Bugelski, B. R. 2, 5, 9, 13, 37

Campbell, Fergus W. 161
Cell Assemblies 40, 145, 146–8, 149, 244–7
Chunking 38
Clowes, Max B. 121, 128
Clustering 38
Cognitive Dissonance 258
Cognitive Psychology 1, 37
Cognitive Representation Problem 1
Configurational Possibility 136
Consciousness 3, 6–7, 47, 60, 221, 223, 234–6
Consensual Corroboration 278–81
Constancy Hypothesis 196–7, 199
Cooper, Lynn A. 63, 91

Depth First Strategy 82

Equipotentiality 158–9

Feigl, H. 229
First Person Immediate Psychological State Reports 254–289
Fodor, J. 55, 91
Fourier Analysis 162
Franks, J. J. 96
Frege, G. 10, 127–130
Fruchter, B. 73

Ganzfeld 258
Gazzaniga, M. A. 95
Galanter, Eugene 159
Gestalt Psychology 126, 178
Gibson, James J. 172–4, 197–8
Groot, de, A. D. 73
Guilford, J. P. 73

Haber, R. N. 149, 150
Haber, R. B. 149
Hayes, P. 121–2, 128, 135–6
Hebb, D. O. 2, 46, 139, 145–7, 152, 213–19, 221–2, 231, 244–6
Helmholtz, von, H. 195–7, 200, 211, 217–18
Henle, Mary 187–193
Herbart, I. 217–218
Heron, W. 146
Hochberg, J. 258
Hologram 159–161, 168
Holonomic Theory 155, 157, 168–72
Hubel, D. H. 218
Humphrey, G. 9, 213, 216

Illocutionary Act 265–6, 271–2
Imagery 1–36, 37–46, 47–71, 73–93, 139, 206, 215–16, 218
  criticisms 12, 18, 48–52
  eidetic 139, 143, 148–50
  explanatory power 3–4
  history 47
  hallucinatory 139, 150–153

hypnagogic 139, 150–153
motor 145, 175–177
non-pictorial 225–6
operational definitions 12–13
retrieval before perception 14–15
rotation 50, 63
visual 161
Implicit Speech 41
Inference
 unconscious 195–212, 217–18
Inferior Temporal Cortex 170
Internal Psychophysical Judgment 64
Introspection 5, 50, 139–141, 155, 213, 231, 253–55, 275–77
Intuition 121
Isomorphism 196

Kant, I. 122
Katz, J. J. 55
Koffka, K. 223–4
Köhler, W. 189, 190, 248–9

Lashley, K. S. 151, 157–8
Leask, J. 149–150
Lewis, D. K. 238
Literals 98–119
Localisation, cerebral 157–161
Locke, Don 238
Locke, John 18, 48
Luminous Finger Technique 74–79, 84

Materialism 228–229, 233
 eliminative 236–8
Maxwell, G. 231
McCarthy, J. 24, 121, 122, 135, 136
McGraw, B. 52
Meehl, P. E. 227–8
Memory
 long term 55, 104–5
 recognition 104
 short term 6, 84, 89–90
Miller, George 159
Mind's Eye 1, 4, 7, 15, 79, 144
Minsky, M. L. 121, 123
Moran, T. P. 73, 87–88, 90, 91
Müller-Lyer Illusion 207, 208, 211

Natsoulas, T. 5, 10
Newell, A. 79, 82, 90
Nuwer, M. 160

Optical Information Processing 160
Organization
 explicate 183–4

hierarchical 27–8
implicate 183–4

Paivio, A. U. 6, 8, 12, 37, 47, 153
Peirce, C. S. 213–14, 216
Perception 51, 92, 142–3, 222
 during occlusion 222–3
 ecological model 172
 epistemic 204
 "grain" 241–9
 qualitative content 223–6
Phenomenal Object 187–193
Post-Synaptic potentials 47–8, 164–8
Predication 101–103
Pribram, K. H. 155, 221, 247–8
Primacy Effects 38
Pritchard, R. M. 146
Protocol Statements 33, 74, 256, 289
Pylyshyn, Z. W. 26, 27, 48, 57, 58, 59, 221

Quintillian 47–8
Quinton, A. 221

Racine, B. 73, 75, 84
Ramus, Peter 53–4
Realism
 constructional 172–3
 direct 172, 179, 187–93
 structural 238–9
Recency Effects 38
Recognition 98, 101, 105–119
Reichenbach, H. 224
Representation
 analogical 52, 60–70, 91, 121, 127–135
 data structure 21–3, 24
 digital 91, 167
 dual-coding 8, 31, 60–70
 information-processing 21–35, 89–93, 96
 modality dependency 95–119
 non-propositional 97, 122
 procedural 23–27, 57
 propositional 7–12, 16, 19–21, 24, 29, 52–60, 97, 119, 136
 prototype 52
Rorty, R. 237–8
Rosch, E. H. 52
Rosenberg, S. 95, 97, 109
Russell, Bertrand 187
Ryle, Gilbert 2, 152, 316, 276

Schematism of the Understanding 206
Seaman, G. 258

INDEX

Sellars, W. 241–3, 246–9
Semantic Distinguishers 56–7
Sensation 142–3, 198–211, 214–15, 276
  projection 179
Shank, R. C. 101
Shepard, R. N. 14, 50, 63, 91
Simon, H. A. 28, 79, 82
Slagle, J. R. 19
Sloman, Aaron 121
Slow Potential Microstructure 167
Smart, J. J. C. 238
Spatial Frequency Analysis 162–8
Spatial Visualisation Tests 31–2, 73–93
Spence, Kenneth 252, 258, 261, 270–2, 274
Sperling, G. 6
Sperry, R. W. 221, 229, 234–5, 236, 241–3
Stimulus Size Estimation 64–70
Strawson, P. F. 206
Structuralism 275–7

Subjectivity 188–190, 230–3

Thom, R. 182
Tolman, E. C. 228
Translations 98
Triebel, W. 258

Validity 122

Watson, J. B. 41, 50
Wiesel, T. N. 218
Winograd, T. 24
Wittgenstein, L. 206
Woodworth, R. S. 143
Workspace
  cognitive 28–9, 30
Würzburg 50

Zimmerman, W. S. 73

THE UNIVERSITY OF WESTERN ONTARIO
SERIES IN PHILOSOPHY OF SCIENCE

A Series of Books on Philosophy of Science, Methodology, and Epistemology
published in connection with
the University of Western Ontario Philosophy of Science Programme

Managing Editor:

J. J. LEACH

Editorial Board:

J. BUB, R. E. BUTTS, W. HARPER, J. HINTIKKA, D. J. HOCKNEY,
C. A. HOOKER, J. NICHOLAS, G. PEARCE

1. J. Leach, R. Butts, and G. Pearce (eds.), *Science, Decision and Value*. Proceedings of the Fifth University of Western Ontario Philosophy Colloquium, 1969. 1973, vii + 213 pp.
2. C. A. Hooker (ed.), *Contemporary Research in the Foundations and Philosophy of Quantum Theory*. Proceedings of a Conference held at the University of Western Ontario, London, Canada, 1973, xx + 385 pp.
3. J. Bub, *The Interpretation of Quantum Mechanics*. 1974, ix + 155 pp.
4. D. Hockney, W. Harper, and B. Freed (eds.), *Contemporary Research in Philosophical Logic and Linguistic Semantics*. Proceedings of a Conference held at the University of Western Ontario, London, Canada. 1975, vii + 332 pp.
5. C. A. Hooker (ed.), *The Logico-Algebraic Approach to Quantum Mechanics*. 1975, xv + 607 pp.
6. W. L. Harper and C. A. Hooker (eds.), *Foundations of Probability Theory, Statistical Inference, and Statistical Theories of Science*, 3 Volumes. Vol. I: *Foundations and Philosophy of Epistemic Applications of Probability Theory*. 1976, xi + 308 pp. Vol. II: *Foundations and Philosophy of Statistical Inference*. 1976, xi + 455 pp. Vol. III: *Foundations and Philosophy of Statistical Theories in the Physical Sciences*. 1976, xii + 241 pp.

DATE DUE

1000003089
HOUGHTON COLLEGE LIBRARY - Houghton, NY
100003089